Introducing CIM for the Smaller Business

P Crumpton

British Library Cataloguing in Publication Data

Crumpton, P
Introducing CIM for the smaller business.
1. Manufacturing Industries. Automation
I. Title
670.427

ISBN 1-85012-810-2

First published in 1992 by:

NCC Blackwell Limited, 108 Cowley Road, Oxford OX4 1JF, England.

Editorial Office: The National Computing Centre Limited, Oxford Road, Manchester M1 7ED, England.

Typeset in 11pt Times Roman by M Wilson, The National Computing Centre Limited, Oxford Road, Manchester M1 7ED; and printed by Hobbs the Printers of Southampton

ISBN 1-85012-810-2

Contents

What the Customer Wanted

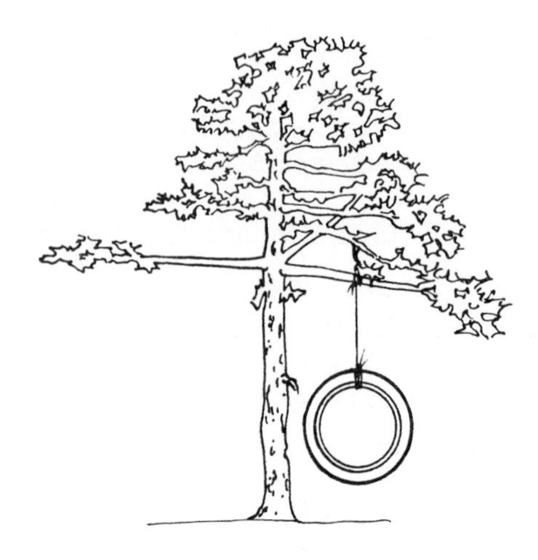

What the Finance Department Budgeted

What the Engineering Department Specified

What was eventually installed

Integration — The avoidance of 'islands of automation' which cannot communicate

First requirement: A shared business objective!

Preface

The writing of this book was prompted by discussions with a number of proprietors and managers of small and medium-size businesses who have realised that the cost of powerful computing has now reduced to the extent that any one of the computing functions of large company Computer Integrated Manufacture should be possible on a desk-top computer. However, they have often been unsure of how to approach the integration of Computer Assisted technologies and have often found the field difficult to read about, since descriptions in the technical press are generally so peppered with acronyms as to be indecipherable.

The chief difficulty has been that, although the computing power to carry out the functions can arrive instantly in a box, any organisation takes time to adjust its structure to suit the new capabilities, especially if the new freedom gives the scope to carry out activities which did not exist in the organisation before.

The larger companies which used mainframe computing to accomplish the functions of computer assisted design and manufacture had the advantage of several years to absorb the stages of steadily advancing automation, changing the organisation gradually. For example, the Austin Motor Company created its own computer-controlled machine (a modified vertical milling machine) in 1959 for use in producing prototype parts, including gear cases for the Mini automatic gearbox. However, CAD was not used until 1968, when Pressed Steel Fisher used a succession of different CAD approaches for curve smoothing in body design, and it was not until 1978 that true CAD/CAM was used for engine prototype parts. Instantaneous on-line, real-time production scheduling was used with the introduction of the Metro in 1980, but it was not until the introduction of a Flexible Manufacturing System for production of the cylinder head of the

Rover 820 engine that the full range of functions now known as CIM could be said to have been adopted at Longbridge.

This book is intended to give some of the historical background to the developments which make up CIM and to describe the reality behind the techniques for the benefit of companies which do not have 25 years to spare and must make a quick assessment of the combination of elements of CIM which will have the greatest impact on the success of their business.

For those requiring to go into further detail for a practical implementation, the companion volume 'A Management Guide to Computer Integrated Manufacture' gives further assistance with case studies and reference material including current communications and data handling aspects of CIM architecture.

Introduction

The Aim of the Book

The aim of this book is to introduce to smaller businesses the application of Computer Integrated Manufacture (CIM), which has been a goal of many large businesses for some years, using powerful and expensive mainframe computing facilities.

The availability of high-power, low-cost desktop computers with suitable software now offers the potential benefits of CIM to the smaller company, but only if its introduction is planned in the context of the business strategy of the company concerned. The aim is to include all areas of a company in the planning approach, so that a full application would lead to the achievement of total business integration.

However, any individual manufacturing company will find that some of the application areas will be more cost-effective than others, with the consequence that a self-financing approach to investment will indicate that these areas would be selected for the initial investment phase. Provided that an appropriate grouping of projects is selected and that sufficient infrastructure is included (such as a communications network) it will then be possible to steadily extend the working subset to a more comprehensive application of CIM, at the same time providing the flexibility of meeting the changing requirements of the business environment.

It must be stated that no book on CIM can provide all the solutions for a do-it-yourself approach to implementing CIM. It will generally be preferable to retain the assistance of consultants who have demonstrable experience in the introduction of the appropriate technology. The reader should, however, be well enough informed having reached the end of this book to select the areas for likely future investment and thus be in a position to evaluate the appropriate areas of assistance such as equipment suppliers, independent consultants, and academic institutions.

The Development of CIM

Computer Integrated Manufacture is the term which has become accepted for an approach linking together a number of independent applications of computers in manufacturing, in such a way that information can be readily shared by, and transferred between, these applications. It is, therefore, more a philosophy of approach than a particular technology, although in practice it depends upon a range of technology applications which have been developed and proven over many years as stand-alone implementations. These individual areas, now generally referred to as '*Islands of Automation*', have given dramatic increases in productivity in many cases, usually by allowing the introduction of an automated approach, together with a degree of flexibility which was not possible before the addition of computer assistance. It was soon recognised, however, that establishing interworking between these *islands* (such as Design, Engineering, Manufacturing Control, Process Planning and Purchasing) was a problem. Transferring information manually gave disadvantages in speed and quality which could nullify the advantages of the automation of the islands themselves. A number of steps were then made towards integrating the islands. (*see* Figure I.1).

The first move towards CIM was to link pairs of applications. It was for technical reasons, rather than any economic motive, that the separate disciplines of *Computer Aided Design* and computer-assisted numerical control were brought together as *Computer Aided Design/Computer Aided Manufacturing (CAD/CAM)*, where the numerical database representing the geometrical design of a component was used to obtain and transform data which will be used to control the drive axis motors of milling machines, lathes and similar equipment. The machines were then used to create the external features of the component itself, which was often of great complexity. These complex designs presented substantial difficulties in ensuring satisfactory quality from manual methods, for example, aircraft aerofoil sections (which needed to be milled from the solid) in the early years of supersonic aircraft production. Also, the data from CAD, with the addition of appropriate tolerance information, was also used to drive *Coordinate Measuring Machines (CMMs)* to check the accuracy of the resulting machined surfaces.

Further Developments

Similarly, other pairs of application areas were linked closely together for specific technical advances, often with economic and management benefits following in succession. Initially, these links were created by the larger user companies themselves, some of whom had developed the application software for their own unique application requirement. Gradually, a more generalised market emerged for the software, which consequently became more general-purpose in character.

As specialist software companies developed their markets and appreciated the needs for integration, more links between software were developed, but often only between modules of software supplied by that vendor. It is only in recent years that it has been recognised by the larger systems suppliers that the requirements of CIM are too great for any individual company to address alone. As a result, there are now regular demonstrations of interworking between vendors and an international emphasis on open standards which will permit all applications to intercommunicate.

Having linked the core engineering functions to the commercial activities of the manufacturing business (such as providing cost estimating information as an output of the initial design process), it has become clear that the benefits to be derived are valuable to businesses operating well outside the original development areas of aerospace and automotive manufacture. Indeed, the inclusion of commercial and distribution activities in the areas addressed by CIM means its adoption by companies with no actual manufacturing activities can bring advantages. In these circumstances, the term 'total business integration' is a more appropriate description for the goal towards which companies in many different fields are striving.

This book deals with the adoption of the elements of CIM in a manner which facilitates their interworking, with particular emphasis on the use of technology to which the smaller company can realistically aspire rather than the mainframe technology.

Throughout the book, reference will be made to a small company which is used as a case study to evaluate the effects of the introduction of CIM techniques. For convenience and to embody as much as possible the typical approaches to CIM, the company chosen is a supplier of production tooling for non-ferrous metal castings used in the motor industry. This enables the results of the practical implementation of CIM to be illustrated in a suitable setting.

Where the requirements for a specialised industry such as a textile manufacturer differ significantly from the example used the features will be compared and contrasted with the 'standard approach' used by the automotive toolmaking company. For a more comprehensive treatment of case studies in different industrial sectors, the reader is recommended to refer to *'Management Guide to CIM for the Smaller Business'*, also published by NCC Blackwell.

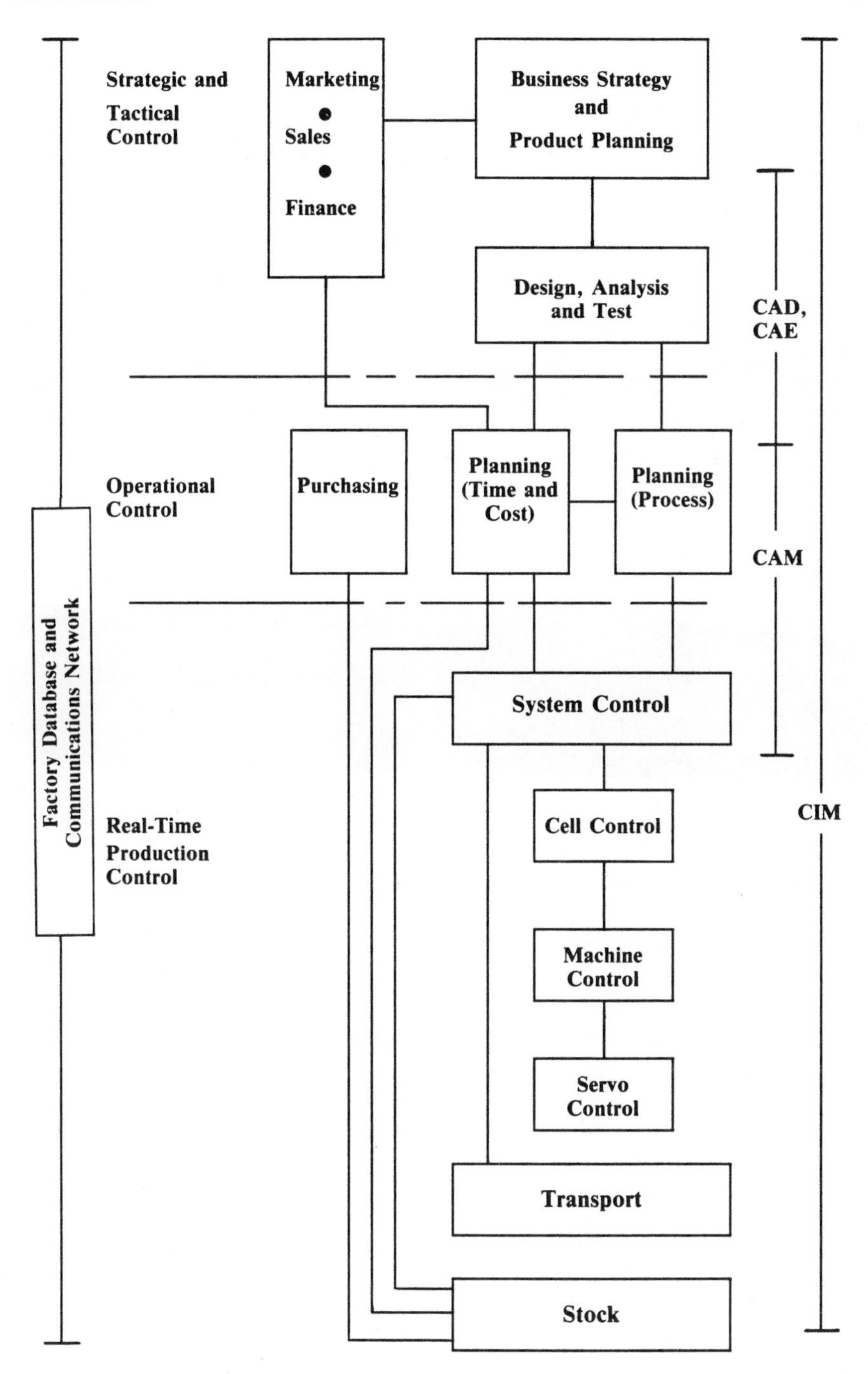

Figure I.1 Reference model for manufacturing control system illustrating scope of CIM

1

Potential of CIM

Introduction

It is a matter of record that the development of Computer Integrated Manufacture within the larger engineering companies has not proceeded to the extent forecast in prediction studies conducted by all major companies carried out in the late 1970s. This is in spite of the subsequent development of technologies and application techniques substantially in advance of those which were predicted at that time. There are a number of reasons put forward as an explanation for this state of affairs; one is the commitment to old systems, representing a high level of investment in unique software, the other is the hardware investment.

For smaller businesses, however, there was previously little incentive to invest in the high degree of computerisation needed for CIM. Therefore, large amounts of capital were not tied up in systems which are acknowledged to be difficult and expensive to integrate. There are now numerous opportunities to benefit from the reduction in the cost of computing power which has occurred in recent years, and build on the lessons to be drawn from the past successes and failures of the larger companies in applying CIM, pointing to a successful prospect for the computer integrated business of the future.

Competitive Advantage or Essential Requirements?

The requirement for business integration amongst the smaller companies in Europe - known as SMEs or small and medium enterprises in the jargon of the European Commission - is recognised as a source of competitive advantage at present and it is soon likely to become requirement without which business credibility, and thus entry to a market, will not be possible. The speed at which CIM moves from behind a competitive weapon to being a required commodity will naturally differ from one industry sector to another.

One of the sectors where the greatest pressure in this area has been felt, has been the suppliers of components and services to the motor industry. Two principal reasons for the pressures here arise from separate areas of the development cycle, design and delivery scheduling. In design, the philosophy of joint development between supplier and principal manufacturer, together with the move towards *simultaneous engineering*, with its iterations between product design and production engineering, require that maximum use is made of Computer Aided Design.

It is natural, therefore, that the large companies which dominate the motor industry should wish their suppliers to use CAD methods, preferably using the same proprietary system as that used by the major company. In this way, any modifications to the component design can be imported to the main database with a minimum of conversation complications.

Although there are means available for conversion from one system to another, for example *Initial Graphics Exchange Specification (IGES)*, DXF, and direct translators (described in more detail in Chapter 8), the repeated use of the transfer for each stage in the development of an engineering design project can cause problems with rounding errors in the software mathematics routines, owing to repeated approximations in the conversion process. Where extreme accuracy is required in the production of complex surfaces this problem is considered so serious that some companies will insist that their suppliers use only the proprietary software used by their own design department. Fortunately, for the small supplier in this position, the proprietary software which was originally only available on costly special-purpose minicomputers is now available for general-purpose workstations and a variety of the software is also available on desktop microcomputers.

The availability of recent developments in the IBM Personal Computer and its 'clones' using the 386 microprocessor with a 387 mathematics co-processor, gives a satisfactory speed of response for the design of medium-sized components, while the recently available 486 chip offers processing speeds and memory-addressing capability in excess of many older minicomputer systems.

Links with EDI

The other principal area in which automotive suppliers are being pressured to conform by the major motor manufacturers, is in the adoption of *Electronic Data Interchange (EDI)* in conjunction with short-cycle delivery scheduling, with the minimum inventory held at the assembly plant (known as *Just-in-Time or JIT*).

This technology can also be used to assist the design process. For example, where a drawing is required as part of the invitation to quote for supplying a part, both the CAD/CAM design data file and the EDI commercial information could be transferred together, probably over the international telephone network.

Developments such as later versions of X.400 electronic messaging with graphics capability will make this more widely available, but at present the practice for EDI and CAD/CAM data transfer is quite different from each other in the nature of the data to be transferred and the probable size and frequency of transmission.

The main advantage of EDI, when used in conjunction with the Just-in-Time philosophy, is avoiding the disadvantage of mountains of paperwork whilst at the same time gaining from the operation with minimal stocks of components, and thus avoiding a large amount of working capital tied up in large stockholding. Hence, a particular supplier might be requested to deliver three times per day instead of the once per month prior to JIT implementation. Without the use of electronic data interchange, each delivery would probably require consignment notes, advice notes and invoices, as well as notification of changes in requirements. All this would have previously been conducted by post, and could result in seriously unbalanced deliveries. The small supplier could therefore be persuaded by a major manufacturer to adopt EDI and he may be supplied with the preferred terminal equipment at a subsidised cost. A further enticement to use the technology may also be provided in the form of an improved cash flow resulting from the quicker invoicing cycle and the possible use of *Electronic Funds Transfer (EFT)*, to make payments to the supplier's bank account.

Modular Approach

Given the ability to design the business requirement for functional modules with integrated communications for the businesses as a whole top-down design (it should be possible for most company operations to implement the modular functions on a piecemeal basis so that each module is capable of operating on its own (bottom-up implementation)). They should preferably be self-financing in the short term. Furthermore, a business system planned in this way is generally capable of maintaining a reasonable level of operation even if one or more modules are disrupted for some reason (a '*robust*' system with '*gentle degradation*'). Later chapters in this book deal with the approach required to plan such a system and the techniques and tools available to help with the implementation of Computer Integrated Manufacture.

2

Computer Equipment and Software

Introduction

One of the factors permitting the adoption of computer integrated business by the smaller company is the ready availability of low-cost computing facilities, with high processing power of the standards of a few years ago. Whether a processing unit is termed a desktop computer, a workstation or a minicomputer, the technological ingredients are likely to be very similar.

A high-speed *Central Processing Unit (CPU), possibly using a single Very Large Scale Integrated circuit chip (VLSI)* will be used to access high-density, high-speed *Random Access Memory (RAM)* chips capable of storing around one million characters of data. The CPU will be supported by other chips dedicated to more specialised tasks, such as mathematical calculations (floating point co-processor) and communications with the outside world (input/output processors). A high-speed I/O processor will be used to link the CPU with *mass storage*, a hard magnetic disk system storing from 10 million to 500 million characters of data, supported by flexible magnetic disk or magnetic tape cartridge systems, for software input and data back-up requirements.

The link between the hardware chosen and the user environment, including the application software and its accessible displays of data, is the operating system. In the course of the advances in the technologies available in hardware, it has been necessary to make parallel developments in the associated opeating systems before the potential of these hardware developments could be fully realised. The operating system of any computer is likely to be a proprietary development.

Consequently, although there is a large measure of standardisation in the system hardware, to assess its power it must be viewed in the context of its operating system environment.

Types of Operating System

One of the more significant differences which can occur between the hardware/operating system combinations which needs consideration, is whether the system is capable of *virtual memory* operation which is to be found on many of the minicomputer systems. Normally only the *RAM* (read and write memory) area of physical memory is directly accessible by the CPU, and consequently, the maximum size of a software program and the data it is directly manipulating is the RAM size (typically 1048576 characters, or 1 Megabyte) minus the space reserved for operating system use (typically 360 Kilobytes).

However, it is possible for one type of operating system to switch this 640 KB of data in and out of the mass storage, usually a hard, or Winchester disk, which contains 40 Megabytes of data. The operating system will be programmed to predict, as far as is possible, which sections of program and data, or 'pages', will be required next, and copy them from the hard disk into RAM, just before the CPU needs to deal with them. Occasionally they will not be there in time, but generally they will, so the system behaves as if the whole mass storage was RAM memory.

Clearly, there could be problems if you need some part of the program in RAM at all times, perhaps to handle an interrupt routine for specialised communications tasks. Consequently, two approaches to operating systems are needed so that more than one operating system is available for both types of approach on microcomputers. Therefore, vendors offer two types of system - *virtual memory* or *real-time* modes. It should be a system design consideration as to which mode is chosen and this may require the use to purchase two separate machines, such as a virtual memory machine for database tasks and a real-time machine for control of factory equipment, even though overall utilisation would indicate that only one machine should be necessary.

It is a feature of many of the more powerful workstations typically using the 32-bit Motorola 68020 or 68030 processor chips, that the package includes a networking capability. With the right level of implementation it may be possible to share data, software and even processing power around the network. In this case the operating system will need to cover the network activities as well as individual applications, and there will be a greater need to obtain a consensus between user requirements than where stand-alone systems use links merely to communicate data. Where portability of a software between users is an important requirement, the UNIX operating system or one of its variants, is a popular choice.

Other operating system considerations are whether the system is capable of *multi-tasking* and whether it can be used in *single-user* or *multi-user* modes. Although the current trend is for tasks to be distributed rather than centralised on one machine (thus obviating the requirement for a multi-user environment) there remains some applications which are still best met by a multi-user system driving VDU terminals. One of these applications is the kind of accounting system needed by a medium-sized company where closely-related tasks (purchase ledger, sales ledger, invoices, credit control) need to be dealt within a secure environment for reasons of commercial confidentiality. Here, the multi-user systems have greater strength than the more commonly available networked stand-alone workstations. A more detailed review of commercial requirements such as *record-locking* and the use of passwords for individual file access is given later in the book.

The company's requirement for multi-tasking needs to be assessed in the light of the particular application software adopted for each of the functions required in the computer integrated business. It will therefore be addressed in each of the following sections as different aspects of CIM are reviewed.

Developments in Microprocessor Chips for Desktop Computers

One of the most notable aspects of the desktop revolution has been the speed with which each generation of microprocessors has been overtaken by a new rush of chips with even more processing power, memory addressing power and cycle speed. The following diagram gives a brief classification of those currently available, with their principal characteristics. (*see* Table 2.1).

Computer Type	Main Processor (16/32 bit registers)		Mathematical Co-Processor (80-bit registers)
'PC' * & 'XT' *	8008*	16-bit processor with 8-bit external data paths	8087
	8086 (used in many PC/XT 'clone' microcomputers	16-bit processor with 16-bit external data paths	
	80186	As 8086 with some other chip functions integrated with main processor	
	(of advantage to the cmputer manufacturer, but not necessarily to the user)		
'AT'	80286	As 8086 with the addition of a memory protection facility to permit advanced memory techniques, if operating system and application software permit.	80287
'386 SX'	80386 SX	32-bit processor (as 80386) but with 16-bit data paths	80387
	80386	32-bit processor with 32-bit data paths	
'486'	80486	As 80386 with additional integration and higher speed.	
	The above microprocessors run at a variety of speeds, the most common being 6, 12, 16, 20, 25, 30 and 33 MHz.		

Table 2.1

Software Choices - Suites and Packages

The software requirement for small-scale CIM systems is, to a great extent, determined by what is available in the marketplace. Whilst it may be possible for larger companies to commission special software for their own requirements, this is unlikely to be feasible for the small-to-medium sized company, except perhaps for tailoring an element of one module, for example, a dedicated post-processor program for a special-purpose numerically-controlled machine. Indeed, it is principally the large market for a relatively standardised element of software which makes CIM possible for the smaller company. The cost of generating a specialist software module for a microprocessor based desktop computer is likely to be at least as great as that for a minicomputer or mainframe-based system which would be prohibitive, if only a few copies were to be sold.

In reviewing the software available for any level of hardware system for your company, one useful indicator is the extent to which the power of the system is utilised effectively. It has been remarked in the previous section that there has been a tremendous speed of development of new generations of microprocessor technology for desktop computing.

The success of each new generation in attracting widespread sales has been influenced to a large extent by the speed at which software is available to run on the new machines. Hence, it has been in the interest of the microprocessor manufacturers to help the software developers in making a new version of their current products as soon as possible to suit the new technology. This is generally achieved by providing a conversion utility as part of a programmer's toolkit for the new hardware and operating system. Existing software is then converted or 'ported' to the new environment by passing it through the utility program, perhaps with additional editing changes for system variables used in a unique way on that particular software.

Unfortunately, this process only makes it possible for the old software to run on the new system, but not to make best use of it. To gain full benefit, many software modules would need to be totally restructured, or at least significantly reworked. Changes such as the amount of memory compared with disk storage are likely to call for a different approach if optimum use is to be made of the newly available system features.

This restructuring has been a characteristic of software releases since the earlier 8-bit systems. Although the Zilog Z80 microprocessor, (with many more instructions available than the Intel 8080) became the most popular chip for business systems, much of the software running on these systems used only the 8080-compatible subset of these instructions, having been developed on the earlier microprocessor.

However, the differences in each generation of chip capability is now even more significant, with the use of 32-bit addressing and high-speed cache memory capability, and it appears to be taking much longer for the software developers to adapt their products for these new capabilities.

One of the areas of CIM in which the use of these capabilities is likely to be of particular significance is that of CAD/CAM software. As discussed in Chapter 4, CAD/CAM is particularly dependent on the availability of a large memory to retain all the elements of a complex design, especially if it is to provide a quick response.

The availability of high-speed hard disks has helped to obviate this requirement, but disk access is still not really quick enough for the purpose, for example, if dynamic viewing, such as rotation on the screen of a complex component is desired, then it is vital that all the component information should be held in memory. Benefits may also be derived for some design tasks by illustrating the movement of elements of articulated assemblies with a degree of animation on the screen, which again requires that the data should be memory-resident rather than disk-based. For these applications, some suppliers of desktop computers will install additional memory for the graphics requirement and also specialist chips which carry out commonly used graphics manipulation tasks as a hardware function, rather than a software program.

So far, for each development of the microprocessor hardware technology, it has been necessary first to wait for operating systems to be capable of handling the new hardware options and then to wait for versions of software which make use of the new hardware/operating system combination. It has been remarked already that it is generally preferable to chose software which best meets the identified requirement and then to select hardware which can best support the chosen software. In the case of CIM for the small company, it is likely that any hardware purchased will be required to support a number of software tasks and a major concern will be establishing compatibility between the versions of software available and a single state of equipment environment, with appropriate version numbers for hardware and the operating system itself. (For the personal computer and its clones, the *Basic Input/Output System (BIOS),* is held in *Read Only Memory* or *ROM,* where the data or program is permanently coded into a chip). A change in version of operating system can therefore be considerably more difficult than exchanging systems disks occasionally.

Although software is usually declared to be 'upwards-compatible', in practice, each generation of new equipment environment has been found to contain at least one major bug (sometimes described as an '*unanticipated system feature*') so that software producers have had to produce adaptations to their previous software releases, which are often available as a free update. If this procedure is anticipated, then due weight should be given in the selection process to the cost of a software maintenance contract and what updates are included as part of this service.

3

Manufacturing Control

Introduction

One of the elements of the computer integrated business regarded as most profitable is that of manufacturing control. In addition to its value as a stand-alone function, it is also an important element of integration as it bridges across a number of areas of industrial administration.

There are a number of systems in use, generally known by a series of acronyms, eg MRP, JIT, OPT, etc. The difference between these approaches is sometimes simply one of the extent to which the system addresses different function requirements of the company, for example, the difference between *Materials Requirements Planning (MRP/MRP I)* and *Manufacturing Resources Planning (MRP II)*. It can also be a difference in philosophy, as MRP and MRP II are traditionally used to control the level of inventory to an optimum level, whilst the aim of the JIT system is to eliminate inventory entirely. Often, the change in philosophy will require adoption by the company as a whole if it is to be effective; for instance the JIT approach cannot function successfully unless comprehensive quality assurance and process control is also in operation.

There is no reason why this requirement should be a problem for the smaller company, since the communications task of ensuring that all departments are aware of the required philosophy is so much easier. There will, however, be a necessary strategic decision stage before deciding which philosophy should be adopted, since it is likely that the smaller company will be dependent on the attitudes of larger customers and suppliers who may not necessarily share the same enthusiasm for more up-to-date approaches.

Available Systems

Among the systems available are: MRP, MRP II, JIT, Kanban, OPT and Kawasaki, which we will now look at more fully.

Material Requirements Planning (MRP)

Material Requirements Planning, now known generally as *MRP* or *MRP I*, was the first approach developed to address the control of inventory by using the computer, to order raw materials and bought-in components in relation to the orders received or forecast, rather than the more usual practice of ordering from stock level indication.

Manual methods of requirements planning were generally too time-consuming to permit this near-JIT approach, so the general practice was to control stock levels on a maximum/minimum stockholding, using a safety stock level to cover usage during the order and supply lead-time period. A Pareto analysis of item value would then indicate the small number of high-value items where the minimum quantity should be stocked, and for which the stock level would be frequently reviewed. Mid-value items would be reviewed at longer intervals and would consequently be ordered in larger quantities. Whilst the low-value items might be purchased only once a year.

There were clearly savings to be made in cost of stockholding if the minor components could be related to the major assemblies for which they were used. This would be particularly useful in an environment where engineering change would result in an accumulation of obsolete stocks if re-ordering proceeded automatically. The early MRP systems were therefore developed from bill-of-materials processors and their most significant process was to convert the single order for a complete product assembly, via the constituent sub-assemblies, components, bought-in items, manufactured details through to raw material requirements, ie the Parts Explosion.

However, it must be remembered at this point that these early software concepts were limited by the power and availability of the computer hardware of the time.

Early implementations were aimed at large organisations, where the dominant concern would often be the determination of the monthly production rate required in production plants given that variability of sales forecasts in seasonal (but predictable) markets and taking into account the levels of inventory held in the organisation's distribution warehouses.

The output of the weekly or monthly computer run, always a batch operation, would be the required production level to ensure 'safe' stock levels in the warehouses and distribution system. In this way customer service was maintained on the assumption that a stock-out situation on any item, in any retail distribution outlet, would result in customer dissatisfaction and loss of brand loyalty to another supplier.

Accordingly, one of the inherent assumptions made by some implementers of the traditional MRP approach has been that the cost of stockholding is low, sometimes treated as negligible, and that all items must be available from stock. We shall see later that some of the more recent concepts take a very different view to this assumption. It may not necessarily follow, however, that such assumptions preclude the modification of the traditional type of MRP software to support more modern concepts and to meet new market conditions which change the company requirements.

There are other characteristics of traditional MRP software which determine its suitability for new requirements, and which deserve consideration when assessing the capabilities of new software against its predecessors. For example, one aspect of traditional MRP software, where the processing run might take place monthly, was that in calculating the production requirement from the sales forecast, the production capacity is assumed to be infinite. No account is taken of such factors as the breakdown of production machinery, as the time as which planning takes place is so far in advance of the planned production period, that it would not be possible to reasonably predict such events in advance. Only such planned changes from normal capacity as the manpower effects of annual holidays, can be taken into account.

Consequently, the central concern of the MRP computer run, the *master production schedule,* is rather a misnomer. It is not suitable for day-to-day production scheduling, but only provides the level of demand for top-level assemblies on which the requirements planning details for components and raw materials can be based.

Although there are varieties of MRP which can take account of feedback of attained production levels and use, these provide compensation for: under or over achieved planned performance, the requirements for the next time period (*closed-loop MRP*), and to take account of current capacity constraints (*finite scheduling*), it is generally considered that it is inappropriate to attempt to use such software for short-term production scheduling. It is usually uneconomical to carry out a full MRP program-run frequently enough (probably every two or three days) to give a good response to changes in circumstances in the production environment.

The alternative is to use the MRP program to generate the necessary works orders to support sales requirements, whilst allowing local production areas the freedom to carry out their own detail scheduling, or shop loading. This makes it possible to take into account such day-to-day factors as sickness, machine breakdowns and overtime working capability.

For this task, a number of low-cost scheduling programs have been developed, which on the whole, are targeted at specific industry requirements, the majority having been originally developed by users to meet their own special needs.

For the small company, where separate works orders are not required, it may be preferable to use only the sales order and stock control elements of traditional MRP software, leaving the scheduling task to a shop-floor based shop loading system; especially if any bill-of-materials information can be provided by links from a CAD/CAM system (*see* Chapter 3).

This approach is also appropriate for real-time control of *flexible manufacturing systems*, where the local scheduling function is known as the *cell controller* and features prominently in both system architecture and shop-floor communications requirements. This also provides the primary channel for management information from unmanned systems.

General Conclusions on Traditional MRP

The typical MRP suite would consist of a number of modules which could be implemented with some degree of independence, though generally with a preferred order of implementation, commencing with relatively passive modules like sales order processing and financial elements. Further modules would be activated as the system became established and sufficient data had been entered and verified to enable additional tasks to be accomplished securely.

In spite of the limitations of traditional MRP described above, there are still important lessons to be learned from the success or failure of earlier systems. Amongst these are the requirement of accurate data and good training.

Data Correctness

It has been said that it was probably early MRP systems which were responsible for the all-pervading computing truism *Garbage In-Garbage Out (GIGO)*. No matter how sophisticated the computer software, it is at the mercy of the data fed into it by the operators and there is also a tendency to regard the computer output instructions as an infallible oracle, even though it is known that some of the input is suspect.

Under these circumstances, there is no substitute for methodical checking of data when setting up the system, together with the use of any *exception reports* for double-checking areas where doubtful indicators are given (such as one or two negative stock values).

Having set up the system with correct data it is necessary to review the system constantly to ensure that data is updated as required. Clearly, all the dynamic information must be updated and housekeeping routines established for clearing such problems as partly-completed batches which were terminated for some reason (such as reduced requirement after batch splitting). If this is not done, there is a danger that exception reports will lose all significance, highlighting batches which, for the original requirement, should have been completed months ago - but which everyone knows are no longer needed.

In addition to this action to keep the expediting priority horizon in the right place, it is also necessary to periodically review what is normally regarded as static data, such as lead times and process times. Unless these are changing (and in a favourable direction!) the company is not progressing and making use of improved methods. Such improvements should be taken into account in the planning procedure if full benefit is to be gained.

Training Requirements

Among the early protagonists of MRP such as *Oliver Wight* and *George Plossl* in the USA (whose writings on the subject are still worthy of study in relation to current implementation), the importance of training was never underestimated. It is likely that the success of any implementation has been in direct proportion to the seriousness with which the company has carried its training requirement. It also has an effect on the data integrity, discussed above, by ensuring that all company personnel have an understanding of the importance of accuracy in the data being fed into the computer files if the output is to be relied upon.

A good training programme will also be invaluable in communicating the overall business objectives so as to help sometimes estranged sections of the company to cooperate more effectively, once important joint objectives can be appreciated.

One of the legacies of the Oliver Wight treatment of MRP is the widely recognised classification of users, combining the extent of coverage of the system used with the effectiveness of implementation. Accurate data represents the fullest and most effective development of materials requirements planning.

MRP II

Manufacturing Resources Planning, or MRP II, builds on the capability of closed-loop MRP systems with the use of additional modules (cost planning and control) to add in other areas of commercial activities, such as accounting.

There is also, generally, a system capability to simulate future possibilities by using multiple sets of data to model alternative probable environments for the organisation to evaluate different courses of action, and so act as a form of decision support tool.

With more information concerning the enterprise available to the user, it becomes feasible to carry out such operations as resource smoothing and more detailed manpower and capacity planning. The term 'rough-cut capacity planning', is more suitable to the needs of MRP but is inappropriate for MRP II and is better replaced for MRP II by a more descriptive term. This would convey the more detailed representation of component flow through the manufacturing facilities, including dependence on key operations, where appropriate.

The name MRP II, was coined by Oliver Wight in the USA to describe a system which would control the operational, engineering and financial resources of a manufacturing company. With this wider definition,

MRP II coincided with the extents required for the whole field of Computer Integrated Manufacture in many medium-sized companies and may even be considered the heart of any CIM implementation; if it can be achieved in practice, including the control of engineering activities. Even if the ideal is not totally fulfilled, there are also practical reasons, (in relation to organisational communications requirements, discussed in detail below), why it is convenient to adopt the computing equipment used for the MRP II task as the hub of the information network; in particular the interchange between commercial and engineering functions and the shop-floor operations. (More detailed cases studies concerning solutions for particular industry sector requirements are given in the publication *'A Management Guide to CIM'*.

For the smaller company, it may be more appropriate to consider the overall requirements for achievement of the aims of MRP II when planning the general architecture but to actually achieve it by separate, linked systems elements, each of which contribute part of the task, rather than use a specific proprietary MRP II suite intended for use by a large organisation. Indeed, many of the large organisations are now making their operating units autonomous enough to consider this approach themselves, using a common system only for periodic, quarterly and end-of-year consolidation of accounting and financial reporting.

Just-in-Time (JIT)

More a philosophy than a system, Just-in-Time found favour on both sides of the Atlantic during the 1980s as the direction in which manufacturing control should develop for most manufacturing company operations. Although a number of companies had managed their operations in this way in earlier years, it was probably the adoption of this approach by the Japanese automobile manufacturers, such as Toyota with their Kanban system (discussed below), which brought it to prominence as an approach to be advocated as a universal panacea.

Despite its success, there are a number of reasons why it cannot be wholeheartedly supported as a basis for Computer Integrated Manufacture. These deserve further discussion to show an insight on the nature of the inter-related aspects of manufacturing functions, which are too easily compartmentalised and labelled.

The fundamental justification of JIT is straightforward to appreciate. Many of the implementations of traditional MRP over the years have been seen to actually increase the level of work-in-progress and general stockholding, rather than diminish it. We have already seen how it is possible for systems, without links to the engineering change process, to perpetuate requirements for stock of obsolete material. In addition, it is an inherent problem with hierarchical, multi-level MRP which has fixed batch sizes, that an order for one additional assembly, requiring only one detail part, can trigger production of whole new batches of sub-assemblies and their constituent details.

At each level, a contingency factor is also likely to be introduced to allow for possible spoilage in transport or quality non-conformance on a percentage basis, so that a wave of production of unnecessary items is propagated throughout the system. This causes a rise in work-in-progress, which apart from consuming excessive working capital, causes physical congestion on the shop-floor. The additional load on bottleneck facilities causes a delay to other items which really are required, resulting in poor delivery performance and customer dissatisfaction.

By starting with the premise that it is possible to reduce finished stocks to zero (with minimum work-in-progress, and with bought-in components delivered exactly when required), JIT offers to eliminate the above problems at source. Any level of manufacture can be treated as an autonomous zone of control, receiving from its suppliers and delivering to its customer. Details are thus only produced when required for delivery as part of a product, which is already covered by a firm order.

Evidently, there are problems with this approach for any industry which is not already highly efficient. First, the lack of stockholding of bought-in components, manufactured details and sub-assemblies means that any quality non-conformance causes immediate disruption of production. Secondly, the manufacture of detail parts 'as required' in small quantities, possibly a batch size of one item only, would be regarded as totally uneconomical to the traditional production engineer, brought up with the concept of an economic batch quantity determined by the set-up time of the operation.

Those companies which do adopt JIT successfully put a great deal of their resources into dealing with these two major problems, and the numerous other minor problems which occur as JIT is implemented. The first problem, that of quality, can be dealt with by adopting process control rather than inspection. Even 100 percent inspection will not prove adequate if the production process is not fundamentally capable. It may prove preferable to use statistical monitoring to keep each process well within the acceptable quality band. If this is not possible on any operation, then engineering resources must be deployed to remedy the situation by modification or by investment, so that quality of output is assured and predictable at all times.

The necessary techniques for statistical analysis have been used regularly in earlier years, the main writings on the subject by *Juran, Deming* and others in the USA in the 1950s, but it was Japanese companies which implemented them wholeheartedly.

More recent additions to this field have been from Japanese writers such as *Taguchi,* who has been particularly influential in the area of adoption of efficient experiment design to identify the problem areas to be eliminated, before acceptable quality with minimum variability can be achieved. As well as in-house improvements, however, it is clearly important that all suppliers can provide the same assurance of good quality.

Similarly, the second problem area, that of set-up times, can often be dealt with provided that sufficient engineering effort can be spared to concentrate on the task. This is more an area for gradual improvement, step by step, but in some industries the improvements have been dramatic. The changing of the massive press dies used for car bodywork panels, for instance, used to take several days, including time for adjustment to ensure good quality panels. The most efficient companies have now reduced this time to a matter of seconds, by using automated tool shuttle slides and ensuring more accurate location.

The degree of change totally alters the economic batch size and also makes it feasible to use high-cost equipment more flexibly, thus amortising the capital cost over a wider range of products.

In addition to these major problems, it is probable that a number of minor problems, previously hidden by the high level of work-in-progress, will become evident. The successful adopters of JIT regard this as an advantage and proceed to deal with the problems with enthusiasm, often forming teams for this purpose from all levels of the organisation and all relevant departmental functions. As the problems are dealt with, the level of work-in-progress can be further reduced and a new set of problems may be revealed, to be dealt with in turn.

From this discussion it will be clear that JIT is not a universal panacea, but can be of great benefit if the organisation has the will to overcome the obstacles and employs suppliers who will also take up the challenge. There are also of course, circumstances where the lead-times inherent in the industry sector concerned preclude the adoption of JIT. For example, one of the most notable must be the specialist brandy industry. In addition to the obvious lead-time considerations when considering increasing production of a commodity which requires maturing for twelve years in cask, at least one producer is dependent on the supply of a particular variety of mature Limousin oak for production of the casks themselves. To increase the supply for current requirements, more trees would need to have been planted two hundred years ago!

Kanban

One of the varieties of JIT production (the title refers to the Japanese name for ticket), which is the authority to produce at any level in the system. Typically, two small flow-pans or pallets of parts are permitted for any individual component. Only when one is consumed is the ticket released to the production area as an authority to produce a further batch. The size of batch, and pallet, is calculated to cover the time required for process set-up, production and transport.

This system, given goodwill and flexibility amongst the personnel concerned, is self-optimising and needs no computer support.

OPT

This title again refers more to a philosophy than a system, although it is also embodied in proprietary software.

The concept, introduced by *Eli Goldratt*, centres on the review of all operations to determine the fundamental operational bottlenecks. Priority in management attention should then be focused on these bottlenecks, on the basis that an hour lost on a bottleneck operation is lost to the whole organisation; whilst an hour saved on a non-bottleneck operation will be of no benefit and may be positively harmful, since it will increase work-in-progress and thus require additional working capital and possibly cause congestion in the workplace.

Other features of this approach are that it should be a requirement that all goods produced must be for an immediate sales requirement, with no place for a finished goods warehouse; and that the aim of the whole organisation is to increase throughput, whilst reducing inventory and costs.

Some aspects of this philosophy, particularly the cost allocation of bottleneck operations, are in conflict with conventional wisdom in factory cost accounting and traditional works management processes. To aid the understanding of the philosophy, Goldratt produced a very readable novel *'The Goal'* dealing with the problems of a US factory plant manager and his progress in changing the fortunes of his doomed plant by adopting the concept of OPT.

Although there are several prominent companies which have claimed immense savings by adopting the rather expensive OPT manufacturing control software, success has only been achieved after a full training programme at all levels of the organisation and a considerable amount of consultancy support.

On the other hand, a number of companies have adopted the central philosophy after reading *'The Goal'* and supplementary material, and have achieved success by utilising more conventional MRP software. This approach should not be undertaken lightly, however, as the fundamental bottlenecks of most organisations are difficult to locate and require the use of capable, discrete event simulation software to model the system dynamics.

For the smaller company, there is now available a lower-cost version of the OPT software.

Kawasaki

This is yet another manufacturing control approach which originated with Japanese motorcycle and machinery manufacturers and which has been promoted in Europe by management consultants. It shares many characteristics with JIT methods such as Kanban.

Selecting Software Suppliers

Having adopted a fundamental approach to manufacturing control for the business, the next step is to select the most suitable software for that approach.

At this stage, it should be pointed out that because certain software has customarily been used to support a particular philosophy of manufacturing control, it does not need to be restricted to that role only.

As mentioned above, there have been successful cases where a company has implemented OPT using elements of a software package which was originally written for MRP. The most important consideration is to determine which functional elements are required for the business and then which is the best means of delivering that performance, giving due regard to the compatibility of the elements selected.

4

Computer-assisted Design, Draughting and Testing

Introduction

If a company is concerned mainly with product development then the computer-aided design function will rival that of manufacturing control as the dominant function of CIM. Indeed, if the company is concerned with producing one-off products or prototypes for high-volume production by others, it is likely that CAD/CAM will be selected as the initial investment area.

A great deal of publicity has been given to this technology over the ease with which the geometrical data from computer-aided design, for example Figures 4.1a-f and 4.2a-f, can be used to provide the numerical coordinates for a computer-controlled machine tool, ie NC Part Program. This publicity has tended to obscure some of the technical developments of *Computer Numerical Control (CNC)*, machinery which are of particular benefit for the smaller installation, and also some of the other benefits of CAD/CAM for the business as a whole.

The Variety of CAD Systems

Before going further, it would be useful to describe the variety of computer-aided design systems and the extent to which they may be expected to cover the engineering design functions and the whole product life-cycle for different industry sectors, such as mechanical engineering, electronics and AEC (Architecture, Civil Engineering and Construction).

SURFACE MACHINING
CHOICE OF CUTTING PATTERN

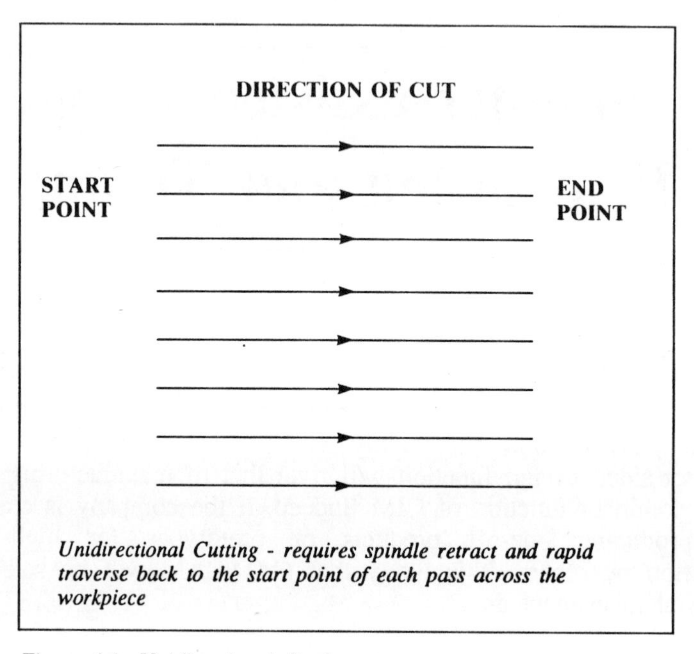

Figure 4.1a Unidirectional Cutting

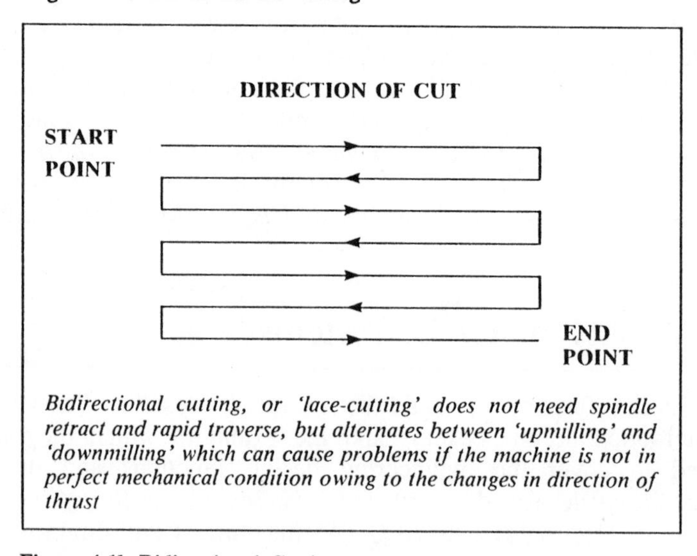

Figure 4.1b Bidirectional Cutting

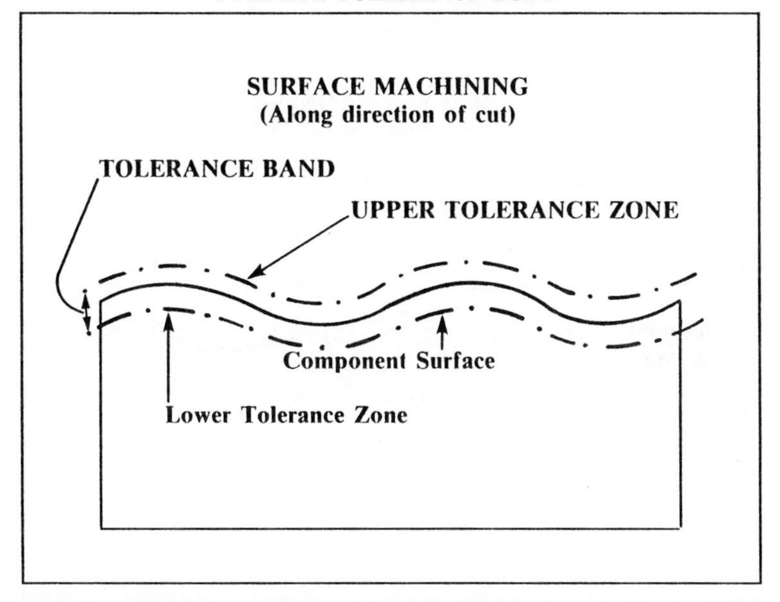

Figure 4.1c Surface Tolerance Zone

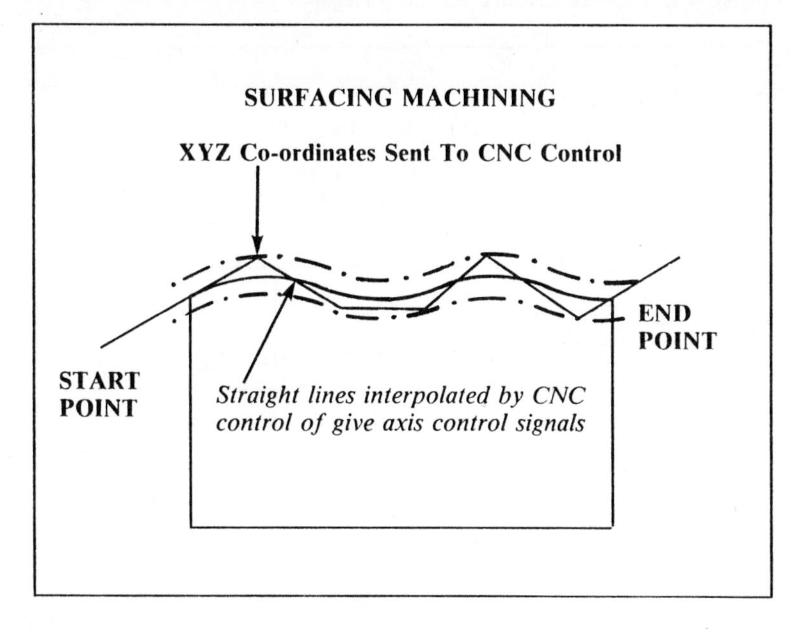

Figure 4.1d Interpolated Tolerance Zone

SURFACE MACHINING — CHOICE OF CUSP HEIGHT AND STEPOVER

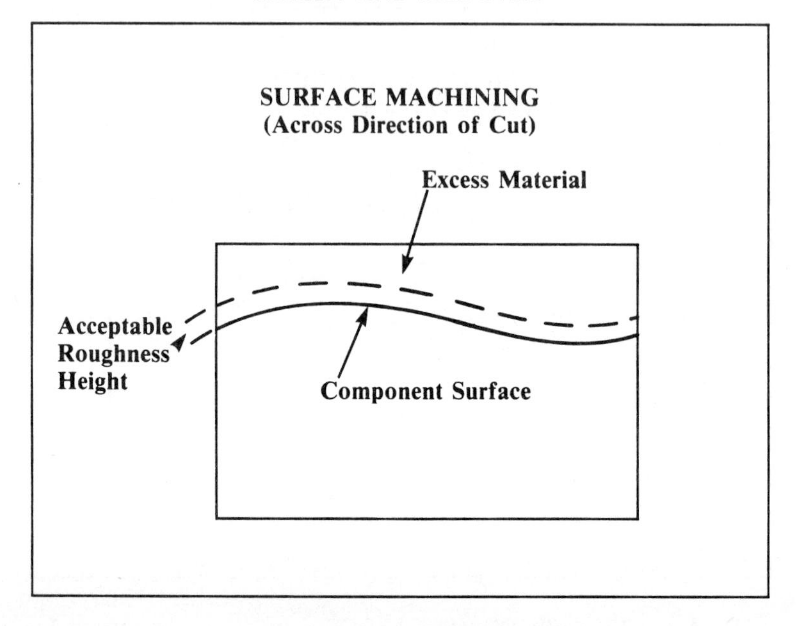

Figure 4.1f Stepover Distance and Cusp Height

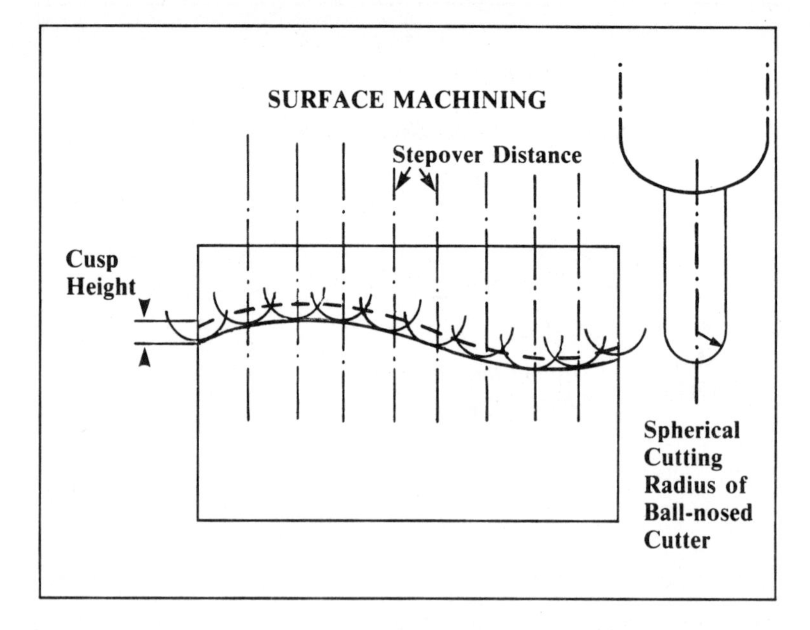

Figure 4.1e Acceptable Roughness Height

CADD

Computer-aided design and computer-aided manufacture might be considered to cover the whole extent of what we now term *Computer Integrated Manufacture*, but the terms CADD and CAD/CAM generally have a more restricted definition arising from the specialised focus of early systems.

Computer-aided design and draughting (CADD) originated with the early two-dimensional draughting systems, where the requirement was to emulate the capability of the manual drawing board, with the conventional three-view representation (plan, side view and elevation) and text addition for dimensions.

As systems gained in capability, it was possible to produce a three-dimensional representation of the part for manipulation by the designer, until it met his requirements for product function (computer-aided design). However, output was still aimed at the production of a paper drawing with two-dimensional views of the part (computer-aided draughting). Provided that exact dimensions were used for the three-dimensional model design, dimensioning of the two-dimensional drawing could be carried out automatically, with a minimum of manual intervention to improve the appearance of the text.

Systems of this type are generally classified by the type of three-dimensional *modeller* used for the design stage. The simplest, the wire frame modeller (Figure 4.3) would model a cube by describing the corners and edges only. No information is held on the nature of the faces of the cube, or whether the interior is solid or hollow.

The surface modellers, on the other hand, in addition to the wire frame information, hold a description of the faces of the cube and permit their modification (such as becoming convex or concave). Since the surfaces are described, it is possible to add information concerning machining paths required to produce the component. It is also possible for a surface modeller, such as *DUCT, Design Using Computer Techniques* (developed by Cambridge University Engineering Department), to indicate the enclosed volume, although this capability is not common.

A *geometric solid modeller*, however, will generally have this capability, and the model representation will indicate whether the cube is solid or hollow. It should also be able to indicate the centre of gravity of the part and by indicating the density of the material, the weight will be calculated. The moment of inertia of the part about any desired axis should also be available.

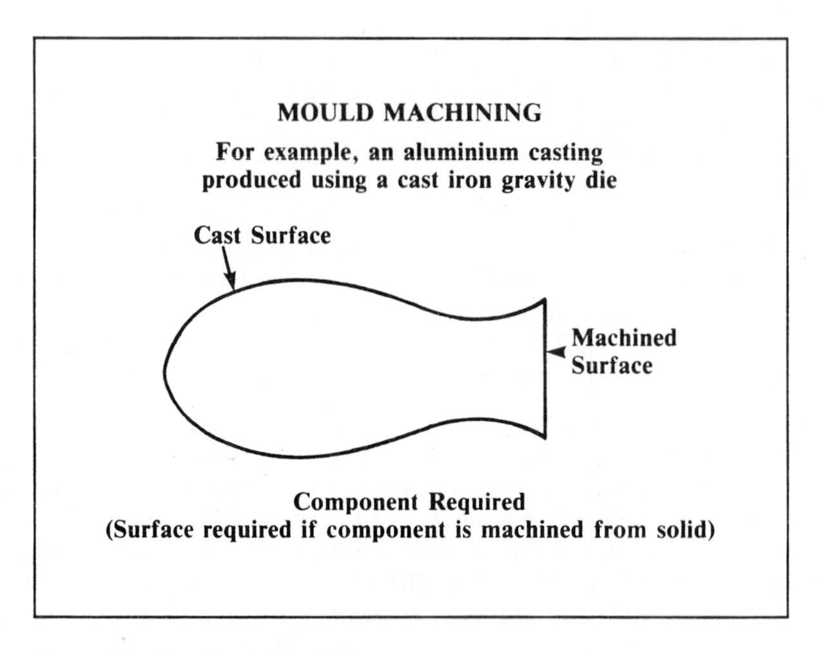

Figure 4.2a Component Required

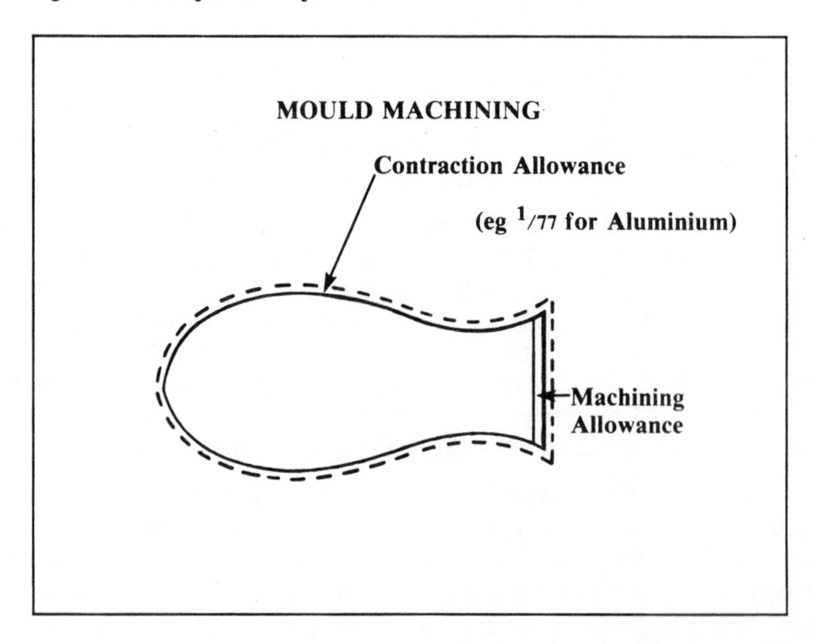

Figure 4.2b Component with Contraction Allowance and Machining Allowance

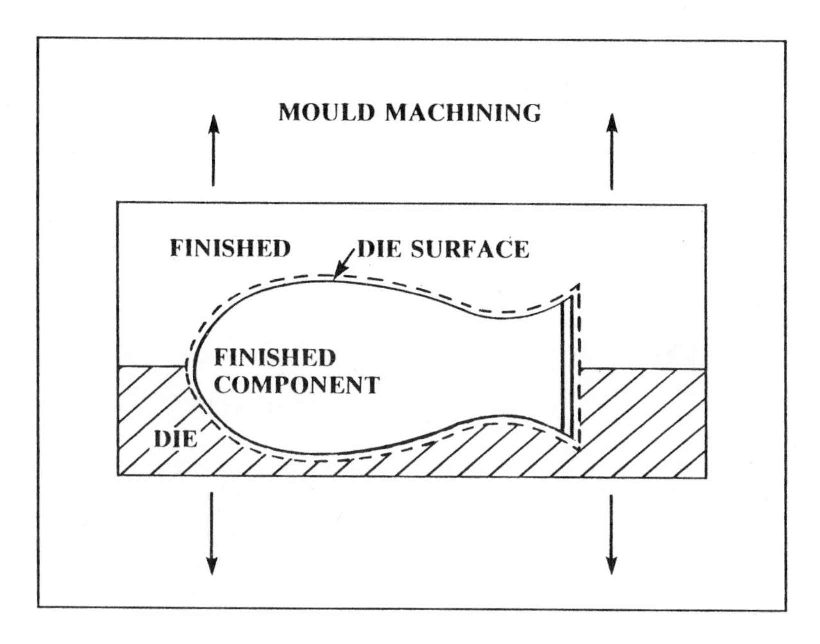

Figure 4.2c Addition of Split Planes to suit Die Separation Method

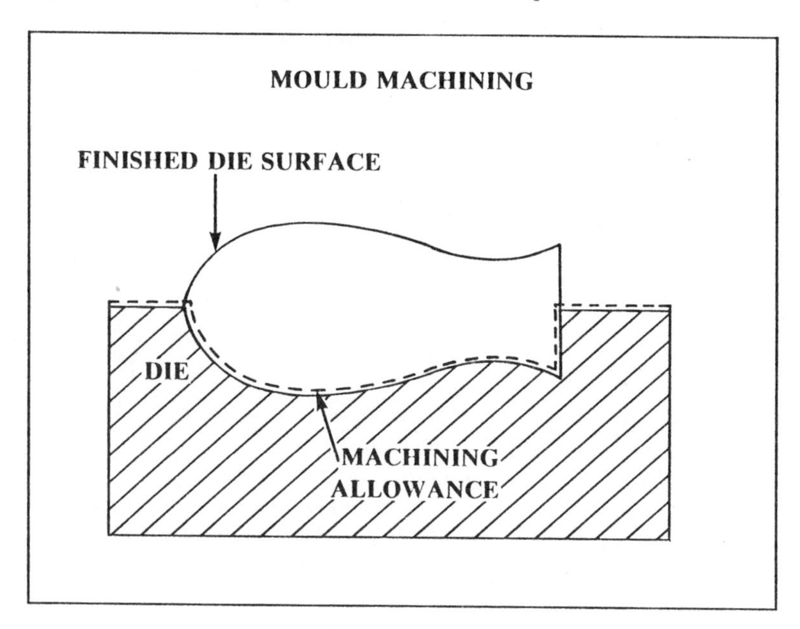

Figure 4.2d Addition of Machining Allowance to Component Surface and Split Planes

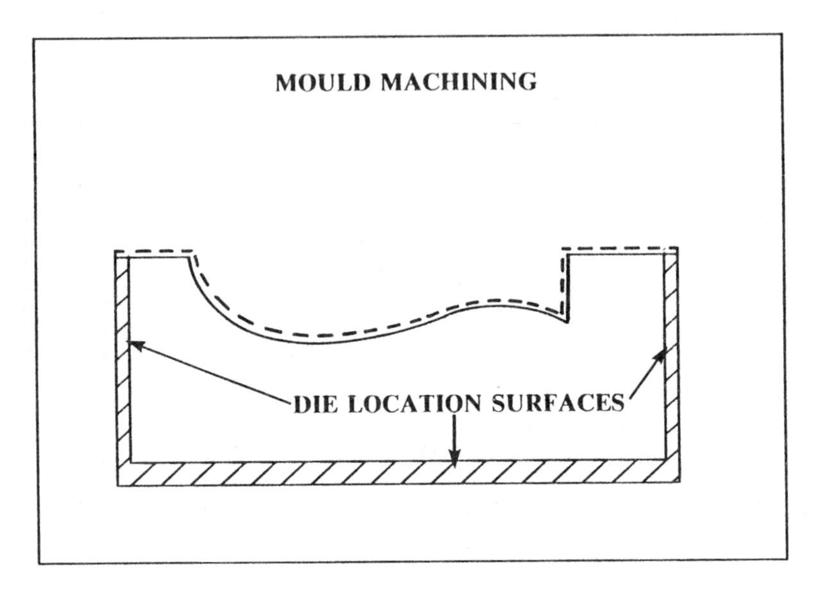

Figure 4.2e Addition of Machining Allowance for Die Location Surfaces

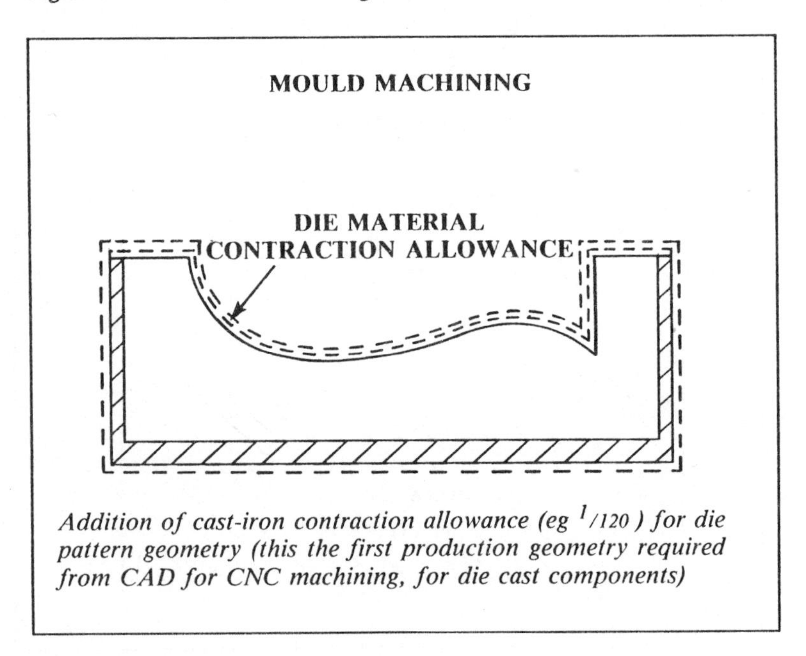

Figure 4.2f Die Material Contraction Allowance

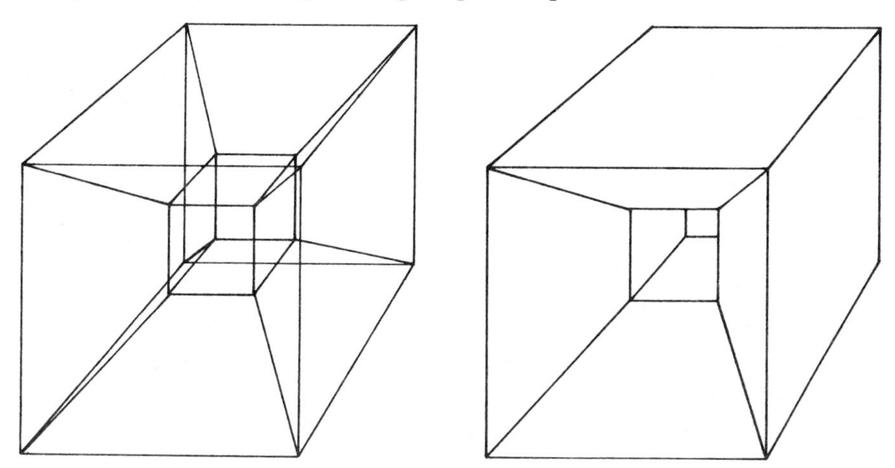

Figure 4.3 Comparison of ambiguous wire-frame representation with ' hidden line removed' representation from a surface or solid modeller

There are two widely-used varieties of solid modeller, the *Boundary Representation type (B. Rep)* and that using *construction solid geometry (CSG)* from geometric primitives such as cones, cylinders, spheres and cubes. It is also possible to combine the capabilities of the surface modeller and the solid modeller.

Whilst computer aided draughting was in its infancy there was a requirement in the aircraft industry for a means of programming support for numerically controlled milling machines to produce the profile-milled panels which were replacing sheet metal airframe construction for supersonic flight. After a concerted development at Massachussetts Institute of Technology (later continued at Illinois), a computer program called *automatically programmed tools (APT)* was completed and put into use. At this time, it used the APT language (based on the general-purpose engineering and scientific language FORTRAN) to describe the component surface and the required cutter path, rather than any graphical representation. The main output of the program was the *CL File*, or cutter location centre-line file, which was then post-processed to a form suitable for a particular machine tool. (Then, more so that now, there were a great variety of combinations of machines and control systems each of which would have their own distinct variation of dimensional/performance details and dialects of language. Consequently, the post-processor programme has usually been specially written for that particular combination. This investment in post-processors is an undeniable force for conservatism amongst existing users and may be responsible for some of the perpetuation of outdated approaches in this area).

As computer-aided design systems developed, it became feasible to use the graphical representation of the component (together with details of the desired cutting tool and preferred direction of cut), to create the equivalent of an APT input file of APT source language commands.

The combination of a CAD system and the APT program was, therefore, able to provide the basis for CAD/CAM for mechanical engineering. Even today, the majority of mainframe and mini-computer-based CAD/CAM systems have the capability to input APT source language statements which allow additions to the graphics-based machining routines to be made. Modern systems also generally have a stage in their process where a CL File is generated, making it possible to use existing post-processors written for an older CAD/CAM system.

It will be clear from this brief description of the development of CAD/CAM that its main focus has been the support of numerically-controlled machine tools and there has been a tacit assumption that any part designed on such a system would either be machined from the solid by NC, later CNC equipment, or produced as a moulding, casting or forging from dies machined by NC methods.

In practice, of course, a decision stage is required after the design approval stage to select the most suitable manufacturing route for each particular component. Alternatively, the decision may be taken to operate a simultaneous engineering exercise combining design and production engineering in an iterative process. This decision process can be assisted in a full Computer Integrated Manufacturing implementation by the use of *computer assisted process planning*, (Chapter 6) as part of linking CADD with manufacturing control. This facility is not available however, in the typical CAD/CAM system.

Engineering Stress Aspects

In addition to using the geometrical model to obtain machining information, it was convenient to use it as the basis for the mesh required as part of the data input of component characteristics for the *finite element, finite difference and boundary element* computer programmes used for stress analysis, thermal modelling and other computer-aided engineering analysis techniques. The combination of computer-aided design synthesis with engineering analysis in this field has become known as *Mechanical Computer-Aided Engineering (MCAE).*

Several of the functions described here have only recently become available on equipment suitable for the smaller company. Although wire frame CAD has been available on desk top microcomputers for some time, the best selling packages have, in the main, been limited to two-dimensional draughting capability.

The availability of cheaper machines using the 386 has changed this position, however, with the porting of surface modelling and machining software similar to that used on the expensive minicomputer systems. Even the geometric solid modeller, which has been widely regarded as too time-consuming and expensive even on minicomputer systems, is now available in a number of forms on the microprocessor.

The more widespread availability of the 486 chip promises to make the solid modeller a practical everyday tool, bringing ancillary benefits, such as colour shading and *ray tracing* to help in the creation of attractive sales brochures. The additional power of the 486 chip will also be welcome in facilitating the analysis of more complex stress analysis problems than it is currently practical to model with the computer-aided engineering software already available on the microcomputer.

Similarly, desk top workstations, typically using Motorola 68020 and 68030 chips, are now available at prices within the range of the smaller company and offer high performance for the MCAE application, especially when networked together.

Where a significant amount of stress analysis is undertaken, the networking approach allows the inclusion of a specialist numerical calculating computer to be attached to the network. This may then be utilised by any user workstation to enhance its performance when heavy '*number-crunching*' is required, leaving the workstation free for the graphical representation of the results.

Electrical and Electronic Engineering

For electrical and electronic engineering, the development of CAD has been somewhat different. Instead of developments in geometrical modelling such as three-dimensional wire frame, surface and solid modelling, the emphasis has been on the use of two-dimensional representation of mask layouts for integrated circuits and printed circuit board design.

In the more sophisticated systems, details of necessary interconnections are input as the basis of the physical layout and varying degrees of computational support are available to use automated routing techniques to optimise this layout, although manual intervention of an experienced engineer is still required for best results.

The simpler systems offer little more than two-dimensional draughting support, although the availability of quick-response colour graphics in the specialised printed circuit board design programs contributes to better utilisation of scarce skills. Both types of program are included in the definition of ECAD or electronics computer-aided design.

Developments from the design stage to the product cycle would include: simulation of the behaviour of the circuit as designed, the use of photographically produced masks (both for integrated circuit diffusion and etching), and the production of printed circuit boards.

The design simulation information also helps in the creation of test programmes suitable for down-loading to automated test equipment (*ATE*). The simpler equipment carries out continuity tests on the printed circuit board (*PCB*) to ensure there are no open or short circuits from etching or plating defects. The more complex ATE tests the behaviour of the board when fully populated with integrated circuits. This field is known as *CADMAT* (computer aided design, manufacture and test), the equivalent of CAD/CAM and MCAE in mechanical engineering.

For the transmission of design data between different systems (where mechanical engineering CAD users would generally use the *initial graphics exchange specification (IGES)* which has a three-dimensional representation capability), the electronics industry uses *electronic design interchange format (EDIF)*. Although only used for two-dimensional representation, EDIF includes powerful facilities for logical structures and schematics, as well as physical layouts of chip masks and printed circuits.

Architecture, Engineering and Construction (AEC)

The AEC field tends to be used to bring together all requirements not covered by mechanical or electronics design. As a result it includes some of the most sophisticated naval architecture and offshore engineering design packages with complex surface capability and hydrodynamics modelling. It also includes the simple two-dimensional programs which many civil engineering/architectural offices find sufficient to provide good productivity improvements for the requirements of traditional office practice.

More sophisticated three-dimensional modelling techniques for modelling of buildings are used, but tend to be developed by the larger users for their own specialised needs. The most common data interchange method in the architectural field is the *DXF* format which, although having some three-dimensional capability, is generally used for two-dimensional layout work. Its simplicity of implementation has encouraged the use of libraries of symbols and standard parts, which can be posted in disk format in lieu of a manufacturer's catalogue.

Localised CAD/CAM

One of the technical possibilities for the operation of numerically controlled machinery in the small toolroom setting is that of a machine tool with a graphics input facility at the operator terminal, resembling that of a computer-aided design system. It may also be possible to define the product geometry and to keep libraries of cutting data for the range of materials commonly used. In this case, many of the requirements of a simple CAD/CAM system may be provided by a single machine tool.

If the majority of machining required is milling of prismatic components, for example, then a machining centre will be the logical equipment for localised CAD/CAM and will be capable of a certain amount of circular work (normally the province of the large vertical lathe) by using either a rotary table or by making use of the circular interpolation facility of the control system to move the X and Y axes of the bed.

Conversely, where the bulk of work is turning, a turning centre with a powered tool turret will be able to carry out work such as spot-facing and drilling, which would otherwise require the use of additional operations on conventional equipment. As a consequence, the dividing line normally drawn by the production engineer between turning and milling work has been blurred by this versatility.

The economies of operation created by latest machines has moved the breakeven lines between conventional and NC production to the extent that NC is now the preferred means for small-to-medium scale production. It is therefore assumed, for this chapter, that a decision has already been made to produce a certain category of component by CNC machining. For full-scale manufacture in large volumes this might not be appropriate, so a later chapter deals with the way in which the process selection decision-making can be supported using an appropriate variety of computer-assisted process planning (CAPP).

An important consideration is at which stage the part programming of components should be carried out. There are now many types of machine tools available that are provided with graphics facilities on their control panel, allowing part geometry to be input directly to the machine and avoiding the need for a separate post-processor program to suit each machine being held at a central CAD/CAM facility. These control systems will often contain technology data details such as the appropriate cutting speeds and feeds for commonly-used materials, allowing good optimisation of the part program. However, the graphics input facilities are generally only suitable for the requirements of part programming, and the same limitations often exist for dedicated desktop part programming systems. Since it has already been decided to produce the component by NC, it would seem possible that the design could be created on the machine graphics facility or on a desktop part programming system located near to the machine tools. However, most designers would require a richer set of operations available to give them the freedom to approach a design task in a creative way. Although the final design could be recreated on a dedicated system, some of the intermediate stages may have no tangible form and could not be expressed by a system based on machining features. These systems often require a very disciplined approach to expressing the part geometry and many creative designers would prefer to return to pencil and paper rather than lose the freedom of a full CAD/CAM system. This will not help us in our aim of attempting to integrate him into the business in a more coordinated way.

Although many CAD/CAM systems are sold on the basis of the direct link between design and manufacture (and there are indeed many benefits such as manufacturing cost reduction and lead-time improvements which can be gained) they should not be gained at the expense of benefits which are of more value to the business as a whole. It will depend upon business circumstances as to which benefits are the more valuable, and a later chapter will discuss the means by which competing solutions, providing these benefits to different degrees, may be compared. For the moment, we will address some of the areas (other than part programming) in which information may be extracted from the computer-aided design process for use in other areas of the business.

Estimating

The use of CAD to provide information for cost estimation on a new product generally requires that more design work be undertaken than is normally economic at this stage, especially when the company does not yet know whether it will be successful in obtaining the order. But for a suitable manufacturing process, such as milling or turning, it must be considered in the same way as any other development cost for a projected new product, as the ability to rapidly model the production cycle can be particularly valuable in assessing speeds, feed and floor-to-floor times for different machining strategies and product configurations, allowing more options to be evaluated in the time available for estimating. In conjunction with a discrete event simulation model of the factory as a whole, it will also allow the estimation of accurate door-to-door throughput times, which may be of great commercial benefit in securing orders, by offering competitive delivery times.

Bill of Materials

For most companies, the most significant step to be made in the progress to a full CIM implementation is linking the design engineering area of the company to the manufacturing control system. The principal means for this is the extraction, from the CAD database, of sufficient information on the components concerned to allow the construction of a *bill of materials* for the relevant product. This can then be used by the manufacturing control system to create the *parts explosion* for the product, together with the complementary routing scheme.

For reasonable results, it will require each of these areas to be fairly well established. It is not until the majority of parts in an assembly are designed by CAD that the transfer software achieves much without a good deal of supplementary information, hand-typed into the system. Nevertheless, the potential benefits are sufficient to merit its provision from the start of the planning process, and also to formulate company standards, so that the level of detail of information held in the CAD system is sufficient to allow the construction of the bill of materials and provide any supplementary information useful to the company, such as preferred suppliers.

Data Extract

Most of the major CAD software vendors either include, or make available as an option, the facility to extract information from the component data file which can then be passed to the manufacturing control system. This information is generally in the form of text notes which may appear on the component drawing, but which may also be placed in a separate data file, linked to the component by name alone.

The data file will not usually be restricted in format, so any ASCII text message can be included for subsequent interpretation by the manufacturing control software. However, there is no standard form in which this information is presented, so it is likely that adjustments in the format will be required to suit the particular manufacturing control software adopted. It is up to the user company concerned to develop a format which will be comprehensive enough for anticipated requirements, but not so onerous that it reduces draughting productivity by insisting on the incorporation of pages of information which may be of doubtful or only transient value.

No generalisations can be made here, as much depends on the allocation of responsibilities within the company; for example, whether there is a combined simultaneous engineering team making both design and manufacturing engineering decisions at the terminal, or whether the CAD operator is restricted to design geometry alone, with a separate production engineering department to decide on component routing and process planning. In either event, the company would be well advised to make use of a consultant experienced in this area before making firm decisions on the selection of systems and their organisation.

Linking to CAPM using CAPP

The traditional CAD/CAM approach effectively assumes that the component concerned will be produced by numerical control machining from a solid billet of metal. This is because the earlier CAD systems were developed in the aerospace industry, with the prime requirement of driving the skin-milling machines to produce the complex surface profiles of wing panels. The link required in this case is for the transfer of surface geometry alone.

The adoption of CAD for more general industrial requirements (such as sheet metalwork and tooling for castings and forgings) requires the bridging of CAD and CAM with a more general link, which can be used to provide information for use in selecting the appropriate production process and routing. The data may be considerably simpler than that required for traditional CAD/CAM; for example, the most significant information about the component itself may be the material from which it is produced and the surface treatment required. This would use a simple text string in the program, rather than a complex geometry file.

There may be decisions involved, however, which go far beyond the normal area of competence defined for the design engineering area in the company organisation. A 'make or buy' decision can often be implicit in the choice of material or process, and the strategic importance to the company may require a full financial analysis of the implications before the design can be released for manufacture. If the decision is to buy, the information required for CAPM will be: supplier identity, 'where used', and quantities per assembly, perhaps with slightly different requirements for service use. However, if the part is to be produced by machining using an in-house DNC or flexible manufacturing facility (*FMS*), the information will need to include the part machining geometry (as produced by traditional CAD/CAM) for down-loading to the appropriate machine when required from the manufacturing system data store at the request of the FMS supervisory controller.

It is also possible to use geometrical data extract facilities to pass details of component features (such as holes and machined surfaces) to *Computer-Aided Process Planning (CAPP)* software so that the decisions on manufacturing methods can be automated. However, the component definition on all proprietary CAD/CAM systems is generally not well enough developed for this purpose at present (with the exception of feature-based solid modellers in the development stage). When feature-based CAD systems which can handle realistic assembly tolerance models become generally available, the use of automated CAPP can move ahead and the transfer of geometric data will be substantial. Until then, the information will be required in the most suitable form for interpretation by the process planning engineer in his calculations of process times and preparation of operator instructions.

Sales Brochures

One of the most rewarding aspects of the use of CAD in the design of a new product is the use of the graphics output to aid the visualisation of the final production item before even a prototype has been made. At the very least, a conventional three-view engineering drawing can be produced to a high standard, creating a good impression of the professional standards of the company when discussing the product with a potential customer.

If the adopted CAD system has a least a three-dimensional wire frame capability, the production of drawings can be extended from the traditional front view, plan and side elevation to include an isometric view which illustrates the appearance of the part when viewed from one corner. This view can be enhanced on some systems by setting vanishing points to create a perspective effect, giving a more realistic appearance to the part as if it is being seen from a particular viewpoint in its normal setting. This is also a very useful feature for use in maintenance brochures and assembly instructions to illustrate the general layout for an operation that may be difficult to communicate in words alone. For the sales brochure requirement, however, the best effect will be gained when a three-dimensional surface or solid modeller is used to generate shaded images of the product. By careful use of lighting direction together with appropriate colours and, in some cases, textures, a very realistic appearance can be obtained. The more recent developments of efficient ray tracing algorithms permits the modelling of complex reflections, for example, the appearance of a glazed ceramic teapot and cups in a glass display case, requiring the creation of the numerous reflected images of the products which would be observed in the real-life setting.

The more sophisticated techniques such as texture modelling and reflection visualisation will clearly require better printing facilities than are generally supplied with CAD systems if full effect is to be gained from the advanced software. It can be possible in some cases to produce satisfactory results from a 35mm photograph of the screen display, but better results are obtained by photoplotter techniques such as Computer Output on Microfilm (COM). These facilities are generally too expensive for in-house installation and would normally be used as a bureau service, communicating the CAD design data by magnetic tape or disk.

The use of colour shaded images has been found particularly useful in the fashion industries of textiles and shoes, where the ability to gain feedback on the likely demand for next year's product is of immense value in allocating production capacity and ordering production tooling.

5

Flexible Manufacturing

Introduction

For the larger manufacturing company, the adoption of flexible facilities has been one of the most difficult areas of investment to justify, since it has generally involved heavy capital outlay in plant and machinery which may show lower productivity than dedicated plant, where the product concerned is assured of a long-term future in production.

A key factor in the adoption of these facilities, however, has been the recognition that markets have become less stable, and that the life-cycle expected for any particular product has been dramatically reduced. Both of these factors require the speed of response which can be provided by programmable automation, which fills in the economic middle ground between the dedicated facilities offering low unit costs and the one-off, toolroom type facilities offering complete flexibility at a high unit cost.

To illustrate the features and benefits of the type of flexible manufacturing generally adopted by larger companies and which have received wide publicity, this chapter distinguishes between the machinery usually associated with flexible manufacturing, and the principles involved.

For the smaller business, it may be beneficial to apply the principles of flexibility even though the usual hardware is not appropriate. A further distinction is made between *adaptable* manufacturing, where the product may be changed economically at infrequent intervals, and true flexible manufacturing, which permits the efficient production of a range of products, where there is no requirement for identical products to be grouped together for manufacture.

Principles of Flexible Manufacturing

There are a number of theoretical principles behind the adoption of flexible manufacturing systems (FMS) which are independent of the technology involved, and which apply to any business in a similar market environment. They are generally not dependent on company size, and indeed, the adoption of FMS by the larger companies has been accompanied by an understanding that they must respond to the marketplace with a speed and flexibility previously expected only of smaller companies.

The chief characteristic of the philosophy is the assessment of the company product requirements as a whole, rather than a number of individual items. An attempt is then made to group the products into families with similar features, since no practical production facility can be totally flexible. At the same time, the families are also rationalised, where possible, to minimise the requirements of vast quantities of production tooling with minor dimensional differences. Given the common requirements of this family of parts, a facility can then be designed to suit the physical requirements for all of the different parts.

If the rationalisation process has been effective, it should be possible, for example, to carry out milling of any component, on any machine, at any time, since a full set of tooling for all parts will be within the capacity of the toolchanger on each machine. Otherwise, a more sophisticated system of tool set preparation and delivery will be required to supplement the workpiece delivery system.

Total Aggregate Volume

Conventional Systems

Having established the requirement for the physical production of the family of parts, the next step is to calculate the number of machines to be purchased. It is here that the major difference in philosophy from that of conventional facility planning is found. For dedicated facilities, where there is no alternative production method, the floor-to-floor time for an operation is calculated and compared with the anticipated peak sales requirement for the product, allowing for utilisation efficiency.

Duplication of a special-purpose machine is expensive, so estimates of required volume will generally be on the high side to avoid the necessity of expansion at a later date. If an additional product is to be introduced in a subsequent year, the peak volume of both products must be allowed for, together with the time lost in the changeover operation.

Flexible Systems

In contrast, the required volume used for planning the FMS can be the aggregate of the products in the next production period (Figure 5.1). If the anticipated total sales volume for the family of parts increases above this level, the capacity of the facility can be increased incrementally, one machine at a time, until the volume requirement is satisfied. Since the machines will be of a standard, multipurpose configuration, they can be bought one at a time if demand drops, which would not be possible with a dedicated special-purpose facility.

The possibility of including prototype parts and producing long-obsolete parts on the same facility, within the same aggregate volume, is particularly valuable to companies with a requirement to innovate, despite wishing to support a loyal customer base for long-established products.

Seasonality

In addition to the flexibility allowed in meeting aggregate sales volume requirements described above, the principles of FMS are also of use in meeting the requirements of components for a number of products with different annual patterns of use due to seasonality of demand. If, for example, the engineering company is manufacturing parts for lawnmowers and for snow-blowers, it is likely that the peak demand for these components (both as complete assemblies and as service parts) is likely to occur at different times in the year. With dedicated facilities, the required capacity would be that of peak demand, unless a high and expensive stock level were to be maintained during most of the year.

With the flexible manufacturing approach, the parts can be produced from the common facility at the appropriate time and the total facility capacity and stockholding kept to a minimum.

Types of Equipment for Flexible Manufacturing

Flexible Machining Systems (FMS) and Cells (FMC)

In the machining environment, the application of computer control to give flexibility to the product mix available from the production unit is likely to be based on CNC machine tools. The programming of these machines will be very similar to any stand-alone application of NC machining, except that the supply of programming is more likely to be via a communications link (DNC) rather than by punched tape or magnetic tape cassette.

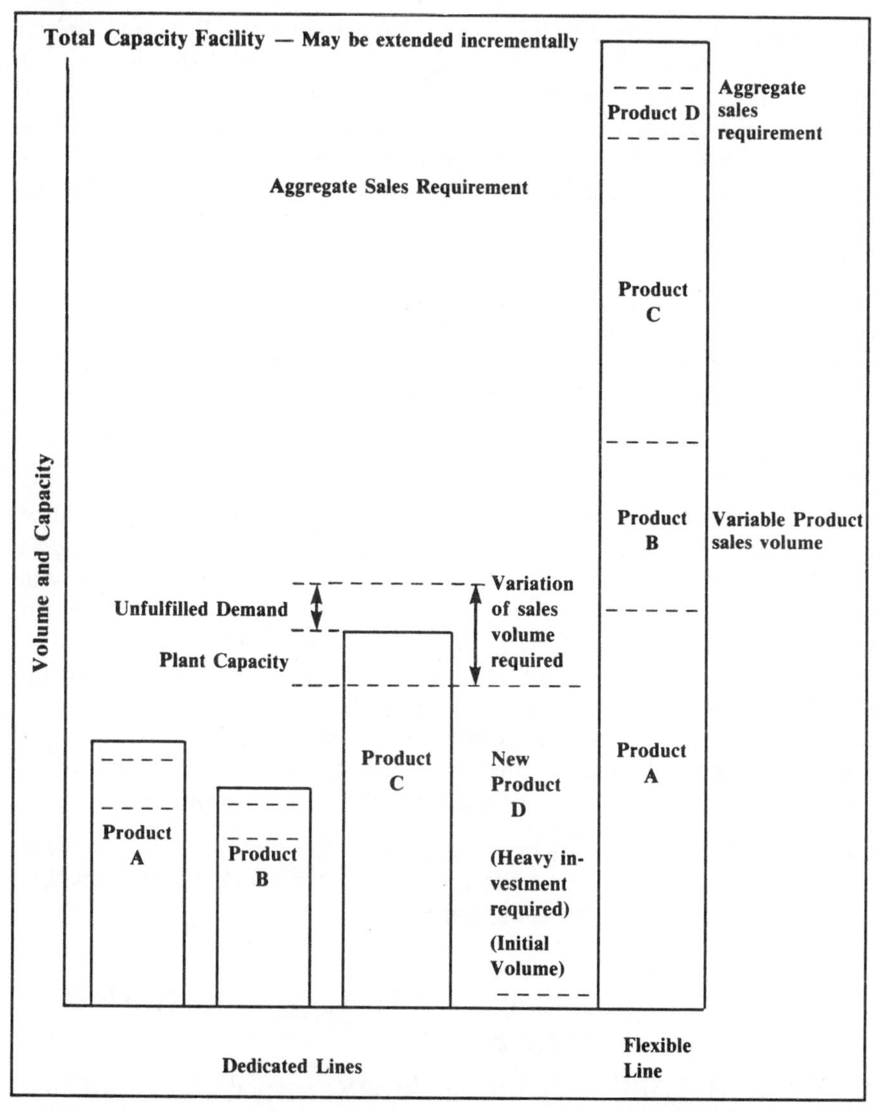

Figure 5.1 The use of flexible manufacturing to improve facility utilisation and reduce investment requirement for new product introduction and where seasonality factors change sales mix.

The main difference in their operation will be the arrangement of work handling for the machines, which, in the case of milling work, will generally mean that component blanks will already be mounted on rigid pallets. Turning machines may also make use of robot loading devices to achieve the same function.

A flexible manufacturing cell of one or two machines may be controlled by a simple supervisory computer or even a fixed program Programmable Logic Controller (PLC), using conventional conveyors between machines.

The full flexible machining system generally requires a programmable transport system (probably with guided trolleys to transport workpieces and tooling), to give flexibility in machine sequencing, as well as product mix. This allows the supervisory control system to mitigate the effects of machine breakdown by using alternate machining sequences to achieve high utilisation of the overall facility.

Flexible Assembly Systems

The development of specialised robot configurations for high-speed assembly operations has given rise to a number of economic applications for high volume manufacture.

The characteristics of final assembly are often unsuitable for automation, as this area is usually most prone to the effects of cumulative errors in the preceding stages of manufacture, and therefore, often requires skilled manual intervention.

Sub-assembly manufacture such as electronic circuit boards, however, is usually more suitable for the application of a high level of automation, where measures can be taken to ensure a high degree of discipline in the supply of quality-assured components.

Flexible Handling Systems

In addition to their use in flexible machining systems, mentioned earlier, flexible handling systems may be of significant benefit even when the majority of operations are manual. For example, the use of wire-guided trolleys for the supply of materials in confectionery factories is long-established, whilst the addition of up-to-date software for warehouse control can add the financial benefits from rapid customer response and minimum stockholding levels to the operating efficiency of the mechanical transport system.

Ancillary Equipment and Support Environment - Gauging, Tooling, Maintenance, Swarf Handling

To fully benefit from the above technologies, particularly FMS, it is essential that a good support environment be created around the system, with well-trained staff and suitable factory services. For example, the work packages may include sets of tooling which will need to be pre-set and calibrated for the specific machining requirements (with any dimension offsets from regrinding being noted against the tool identity), for communication to the machine control system at the appropriate time.

In any large scale operation it is likely that a full tool management system will be required to maintain the records efficiently.

For all NC applications, the question of swarf handling is by no means trivial. Process planning decisions tend to move in the direction of production of components by machining from the solid, instead of minimal machining from a casting or forging. With the high metal removal rates of modern high-power CNC machines that are fitted with suitable tooling, the generation of large amount of swarf can be a limiting factor on output unless swarf processing facilities can meet the demand. In addition, the possibilities of using the flexibility of FMS to machine components of dissimilar metals, on the same day, or even in a single setting, need careful consideration of swarf separation techniques if full economic return is to be made from disposal of high-value non-ferrous swarf.

Implementation of Flexible Manufacturing

For most small companies, the type of flexible manufacturing system which has gained the most publicity, ie the 'lights-out' running capability of highly automated plants with numerically-controlled machines and robots, and with material and tooling delivered by automatic guided vehicles controlled by a central computer for overnight unmanned operation, is likely to be too ambitious an undertaking - although it has been done successfully.

The fundamental characteristics of FMS are, however, applicable to most manufacturing companies and elements of FMS technology are likely to be appropriate. First we will look briefly at the business advantages and then the individual technologies which make up FMS.

The background to the development of the first flexible machining system, often by machinery manufacturers for their own products in the first instance, was initiated by the wish to use the programmability of numerical control machining with the higher throughput possible by transfer line technology. NC machining had improved the economics of the cutting operations for small-batch production by eliminating the need for special fixtures and dramatically reducing the set-up time and cost for individual operations. It had become clear, however, that there was a limit to the large scale application of the technology, particularly for medium-to-high volume production requirements.

By contrast, transfer machining lines used by the automotive industry for components such as engine blocks and cylinder heads, although requiring high investment for only one product, used a multiplicity of machining stations each fitted with multi-spindle drilling and tapping heads. By splitting up the longer operations among a number of machining stations, a cycle time of under one minute per component was obtained, rather than the several hours required for the single-spindle NC machine.

It was also possible to include blow-out stations after hole drilling, to ensure that swarf was removed and that holes were to the correct depth before the next operation. Subsequent developments increased the use of gauging equipment along the transfer line and the use of sensors such as drill breakage detectors, which could automatically stop the line if a problem occurred and thus minimised still further the manual input required to achieve high production rates.

The flexible system was therefore developed so as to fill the gap between individual NC machines and the high-capital cost, low unit cost transfer machine technology, as shown in Figure 5.2. The first successful example was developed by Dr Theo Williamson at the Molins company, manufacturers of high-speed packaging machinery. The concept was for a limited number of specialised NC machining centres for 3-axis and 5-axis milling, each of which could carry out its operations on any of the parts required. Each machine was fitted with transfer slides for components and for tooling, both of which were arranged on standard-sized metal pallets for each of handling. The slides of individual machines could be con-

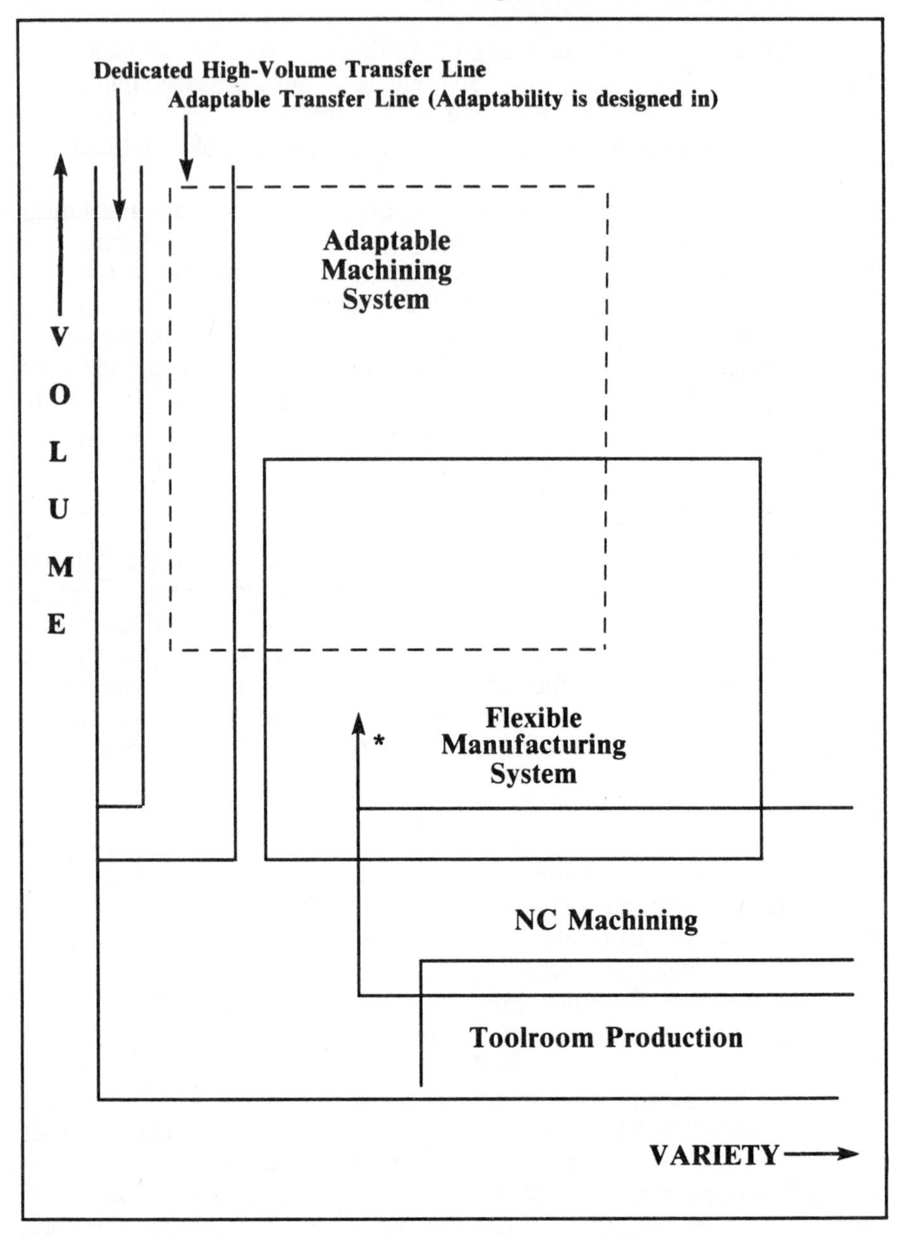

Figure 5.2 Flexible and adaptable machining systems: their place in the technoeconomic spectrum.

* NC machines with addition of tool and workpiece storage facilities.

nected together to form a complete handling system, and automated parts and tool stores (similar in configuration to the machines themselves), could be included in the line to improve efficiency if any machine required maintenance.

Under computer control, the handling system with on-line stores provided for unmanned overnight production, with the manual operations, such as attaching component blanks to the pallets and unfastening the completed components, to be carried out during the daytime shift. This arrangement clearly allowed uninterrupted working 24 hours per day for the line as a whole, and the system was accordingly known as *Molins System 24*.

The special economic features of this approach, however, made it unsuitable for general application. For example, efficient cycle times were obtained by restricting the material to aluminium alloy and utilising the very high cutting speeds which this material permits, using turbine-driven spindles on some of the specialised machines.

The geometrical envelope of the components allowed on the handling pallets were also quite small, but the size and material proved suitable for packaging machine requirements. Other companies with similar requirements, such as aerospace and computer components, are thus, still using Molins System 24 machines over 20 years later.

It is worthwhile reviewing the limitations of this early system, for its characteristics offer an insight into the behaviour of all flexible manufacturing systems. For example, the degree of flexibility, in particular, of any FMS system is likely to be constrained to a family of components within a certain size range and produced from a restricted range of materials. Any decision on adoption of such a system should take place only after a wide-ranging product strategy study to ensure that these constraints do not preclude possible product and market development opportunities. In a more positive vein, there may be an opportunity to rationalise the tooling requirements for an existing range of products for FMS production and to use more effective material utilisation, such as close-tolerance casting and forging, to minimise the machining requirement and improve the economics of single-spindle machining.

Adaptable Machining Systems

A variation of the conventional FMS is the use of machines which can utilise multi-spindle machining heads, either turret-mounted or with a shuttle change mechanism, together with a limited amount of programmable axis control. Some dedicated investment is required, since the hole centres of the multi-spindle heads will only suit one component, or family of components, but the machine will provide for higher production rates than is possible with a single-spindle machining centre. A combination of head-changing machines together with a number of machining centres (with flexible transport in the form of wire-guided or dead-reckoning automated guided vehicles carrying pallets of components), offers the most effective means of producing a relatively stable family of medium-volume machined components, since the machining centres can cater for minor modifications or special customer requirements. In conjunction with real-time control of the flexible transport system and down-loading of part programs from the scheduling system, FMSs can also provide greater reliability for the system by taking over the duties of any of the high-production head-changing machines in the event of breakdown.

Elements of the Technology of FMS

Direct and Distributed Numerical Control (DNC)

The facility available for down-loading programs to individual machines, be they CNC milling machines, lathes, robot or coordinate measuring machines, can be instrumental as an intermediate stage towards full FMS which is now known as the DNC system. There has been a number of approaches over the years which have used this acronym, so it is worth discussing to outline the different versions of DNC, each of which are appropriate, under suitable circumstances, to the small company requirement with current low-cost computing equipment.

Direct Numerical Control

The first use of DNC (with the meaning of direct numerical control), was an early approach to replace hard-wired numerical control systems on individual machines with the flexibility and computational power of the digital computer. At that time, the available computing equipment was large and required a dedicated installation environment, so the preferred solution was direct connection from the central computer to the individual axis drive controller at each machine.

The part program, probably post-processed from an APT CL File on the same computer, would contain individual axis dimensions fed into the machine controller step-by-step at the same speed as the machine's paper tape reader would have operated.

These early machines, if of the point-to-point control variety, would only contain sufficient memory for all the coordinates required for one position. If they were of the contouring variety, there would only be sufficient memory for the interpolation of the cutter position between the fixed data points of the cutter path. It was therefore, necessary for each machine to be continuously connected to the central computer throughout the cutting operation, as even momentary disconnection would result in loss of critical data.

If the absolute system of positioning was in use, a short interruption might lead to a missing hole, but if incremental positioning was in operation, the complete frame of reference would be disrupted for the remainder of that job. As well as this inherent latent unreliability, the continual communications requirement was found to impose an uneconomic overhead on the computer operation, with an expensive resource being unavailable for more demanding computational tasks whilst machines were in operation.

It is not surprising, then, that this system fell from favour, However, the principle is still of use for small-scale applications where it can be very cost-effective to use a desk top computer to drive an axis controller for servo drives, or to drive stepping motors direct. A typical example might be a special-purpose two-axis machine with a high-speed routing head used to profile-cut wooden or plastic panels.

Distributed Numerical Control

The next appearance of DNC is as distributed numerical control, following the development of machines where the hard-wired electronics of the control system had been replaced by a general-purpose computer in each machine tool, programmed to have the functionality of the hard-wired control. Apart from the flexibility of reprogramming for new functions, which was of clear benefit to the machine tool builder, the availability of a certain amount of uncommitted memory meant that the *computer numerical control (CNC)* machine could store the entire part program for a point-to-point machining job, with having to read a punched paper tape each time a component was produced.

By connecting to a central computer which could hold the part program files, the equivalent data to that on the punched tape or '*tape image file*', could be transmitted to the machine. Since it could be stored in memory, there was no need to occupy the central computer during a machining operation, as a program could be transmitted at high speed at a convenient time and the communications facility would not be required again until the machining of the whole batch of parts had been completed.

It should be mentioned at this stage that although the typical CNC controller has a capability for both point-to-point and contouring operation, the memory available for part program storage is often only around 10 Kbytes. Whilst this may be adequate for more than one point-to-point program at a time, a complex surface machining program with a fine cutter step-over may well require over 1 Mbyte. Under these circumstances, the communications process has to be continuous, and the central computer is occupied in a similar manner to direct numerical control, except that larger sections of program can be transmitted at a time, and there should be a better integrity of data transmission. Data communication considerations are dealt with more fully in Chapter 7.

Adaptable Manufacturing

In contrast to the total programmability requirement for flexible manufacturing, adaptable manufacturing is an efficient manufacturing facility which can be economically changed to a new product at infrequent intervals. It may well be appropriate to have some features programmable, for example drill depth settings, as there will be operational advantages over mechanical limit switches and the possibility of incorporating adaptive control, optimising the feed rate for the condition of the material being worked. It may also be economic to use robot loading devices. In general, however, the adaptability will be gained by using modular machine construction, with a heavy emphasis on mechanical design ingenuity to permit the interchangeability of quite different machine elements.

6

Linking CAD with Manufacturing Control

Introduction

The primary development in the merging together of the islands of automation to produce CIM is generally the linking of computer-aided design with manufacturing control, known as computer assisted production management. Since it is difficult for a generalised description to be representative of all varieties of industry and commerce, the example here will concentrate on the requirements of a small metal-working company with design responsibility for some products, and which fills surplus machining capacity with subcontracting work for a major manufacturer. Some of the significant differences for other specific industry sectors will be reviewed later.

Holding Information

Although future systems may well use the Product Design Database as the main repository for all information concerning their manufactured products (including material suppliers, libraries of standard parts and 'where used' information, which show how many occurrences of a common component may exist in the company products), in most cases, computer-aided design is at too limited a stage of implementation to make this feasible. Therefore, the assumption must be made that the main organisational record of company products will be held on the manufacturing control and financial systems. The one-way linkage from CAD to manufacturing control will therefore contain details of individual assemblies, with a subset of the CAD information being transferred, as required, for the tasks in hand.

Linking Functions

There are two established methods of linking the functions for MRP requirements. The simplest is that the assembly *bill of materials* for the product is transferred from the CAD system to the MRP data files, generally in the form of an unstructured ASCII file. This is accepted by the MRP package and converted into the form which will be required at program run-time by a utility programme supplied with the package. However, this had not given many benefits of integration beyond avoiding having to type in the *bill of materials* manually.

Computer Aided Process Planning (CAPP)

A more effective approach to this link requires taking a wider view of the business requirements, in particular, the probable need to provide realistic cost estimates of new designs (whether for internal project control purposes or as a basis for quotations for external customers). This makes better use of information routinely available from many CAD/CAM systems, via a type of software known as *computer-aided process planning*, or *CAPP*.

There are three varieties of CAPP in common use: *generative, constructive* and *variational.* Each have their advantages and some software packages will permit the use of all three approaches.

Generative process planning

Generative process planning offers the greatest power when dealing with a totally new product, and will make most use of the information held by a 3-dimensional computer-aided design system, particularly the form of product description held by a geometric solid modeller. The principle is that the software will analyse the manufacturing requirements for any component and determine the optimum operation sequence. For a prismatic part, such as a rectangular block of aluminium with drilled and threaded holes, the software would assume that a milling machine of a certain size would be required. The size of holes would determine the drilling sequence required, whilst the size tolerance would indicate whether an additional reaming operation would be necessary.

If a complex surface milling cut was required for any feature, the time required could be obtained from the NC programming module of the CAD/CAM system used. For a rotational component, the overall dimensions would indicate what size and power lathe would be required for the turning operations and inspection of the part's other features, such as cross-holes, which, if present, would indicate whether a lathe with powered turret would be necessary, or if an additional machine would be required for the secondary operation. Again, information could be obtained from the CAD/CAM system machining routines as to the time required for roughing and finishing cuts to produce the required component profile from a given blank size of the specific raw material.

Unfortunately, even the most expensive of current CAD/CAM systems are not yet able to provide a comprehensive enough component description to permit effective generative process planning. A great deal of manual intervention is still required. Research developments in this field are progressing, however, and the new generation of feature-based CAD systems may well encourage this work to speedy fruition. Until then, one must ensure the availability of an experienced engineer wherever this approach is used. For example, in the case of the aluminium part considered above, there may be a certain production volume at which it is more economic to produce as a casting with threaded holes, or the assembly requirement may permit self-tapping fixings. These possibilities will be beyond the experience of the CAPP program and will therefore not be considered unless manual intervention is made.

Variational

Variational process planning is more limited and may be more straightforward to implement for a company in which products generally consist of similar types of components and for which new products are judged likely to consist of modified and developed versions of similar components. Essentially, historical information is used to determine time estimates for a component operational sequence, as an assistance to a manual planning process. The resulting sequence, with operation times, may then be passed to the routing module of the MRP system.

Constructive CAPP

Constructive CAPP is effectively a version of variational process planning intended to give some of the benefits of the generative process planning approach. Several proprietary systems are available to carry out this task.

Information Systems - SFDC and SCADA

No integrated system can give comprehensive information about the progress of activities in the business unless it is fed with data in a timely manner, with the appropriate contextual clues to enable the computer to convert the steam of data into useful information when required. One of the reasons for the slow acceptance of computer support for manufacturing amongst the larger companies was the preponderance of batch-oriented mainframe systems with inadequate data acquisition facilities. This led to a response (around two week later than the request), in the form of a pile of line-printer output, several inches thick, which gave an intimidating task for the poor line manager who urgently needed the information for a crucial decision. If the integrated business system is to be relied upon for effective decision support, it is clearly vital that the requirements of the various functional departments are borne in mind when deciding on the nature and extent of the data-gathering systems which the business will require.

Two acronyms in common use describe varieties of data acquisition systems. Although the precise definition, and consequently the extent of their perceived range, will not necessarily be agreed by all authorities, and probably not by any two system vendors, they represent convenient groupings of data acquisition subsystems following two distinct approaches.

Shop Floor Data Collection

Shop floor data collection, or SFDC, represents the approach taken for data collection in manufacturing industries concerned primarily with discrete parts manufacture. In highly automated machining facilities it may be possible to connect on-line, real-time data input to information interfaces on the machines themselves thus allowing immediate response to an urgent requirement on an unmanned machine. More usually, however, the term will refer to the use of bar-code readers or similar scanning devices to take a manual data input from a job-card, together with operator time and attendance information, to permit control of batch manufacture. Its prime role will be as an input to the manufacturing control system.

Supervisory Control and Data Acquisition

Supervisory control and data acquisition (SCADA), on the other hand, will generally be used in a process control environment. Since one of its functions can be to directly control both batch and continuous processes, the time response is generally more immediate than would be required for SFDC. Indeed, as well as the supervisory aspects, it may be appropriate to use the system for direct control of the process rather than use local controllers which report back to the supervisory system. In this case, where perhaps functions were once controlled by three-term controllers (PID or proportional, integral and derivative) the time response may need to be particularly speedy, and more specialised communications solutions will be required.

Determining Needs

Having made the choice of the SFDC approach or the use of SCADA, there remains the task of assessing the extent to which automated input should be used to drive the manufacturing control system. As we have seen earlier, there may be a strong case for limiting the use of an MRP system to that of sales forecasting and requirements planning over a fairly long timescale, such as two or three months. Clearly it would be inappropriate in this context to link in the minute-to-minute, or even microsecond-to-microsecond information from a process control system. However, the real-time scheduling system required for control of a flexible manufacturing system will rarely give of its best unless it is provided with timely reports of the status of each of the machines under its control, so that instructions passed to the transport system, perhaps using automated guided vehicles, can then take account of actual or incipient fault conditions in the production equipment.

Computer Assisted Distribution

Two of the techniques offered by the new generation of hardware and software can be combined to give useful opportunities to the distribution company (or the distribution function of a manufacturing company) which has its own transport department.

The opportunity arising from the widespread adoption of electronic data interchange (EDI), in the context of linked business systems for such operations as invoicing and funds transfer has already been discussed.

Where a number of the principal companies in any industry sectors adopt this approach, it becomes feasible for the smaller supplier in this industry to not only assimilate the use of this technology (this may become a trade requirement in any case), but to go further and develop a strategic advantage from its use.

By linking the demand for information from the EDI system with one of the software packages now available for transport scheduling and optimisation, it becomes feasible to offer a much improved service for immediate delivery from stock or from a flexible manufacturing facility. Such a software package will generally include a traffic routing model based on national maps to enable the most economic route to be generated, taking note of any long-term traffic congestion problems. This gives the driver a hard copy of the preferred route, together with any other delivery documentation required.

Further developments of this approach will undoubtedly include the use of a radio receiver in the driver's cab to give a display of the next section of his route, modified if necessary to take account of the latest traffic problems. Systems of this type are currently undergoing trial operation, mainly intended to avoid congested points in large cities, but their more general availability will herald a further revolution for the distribution industry when linked to the customer in the most direct way.

7

Communications and Architectures

Introduction

These two subjects are addressed in combination as the availability of communications solutions is likely to control the selection of the ideal systems architecture for the organisation. It is strongly recommended that the long-term architecture requirements are developed in conjunction with the business strategy for the appropriate time horizon, before any detailed project implementation plans are considered and any consequent investment takes place. Once the architecture has been reviewed, individual project plans can be evaluated from the point of view of their strategic importance, as well as their cost-effectiveness as stand-alone project implementations. This chapter will deal with the architectural factors; the business strategy and financial evaluation aspects are covered in Chapter 9.

Communications Requirements

First the broad areas of communications requirements are:

- office areas;
- shop-floor communications;
- process control.

Office Areas

The engineering and commercial office environments are broadly similar in nature and the preferred solution can be common to both; although the type of data transferred and the nature of the data transaction may be very different.

The standard solution is the use of baseband coaxial cable transmission using the *technical and office protocol (TOP)* a variety of the ISO *Open systems interconnection (OSI)* recommendations originally developed by the Boeing company for their aerospace design office requirements. TOP also has an equivalent in the variant of OSI recommended by European and US governments for administration tasks (*GOSIP, or government OSI profile*).

Shop-Floor Communications

For the shop-floor data communication requirement the general recommendation has been the *manufacturing automation protocol (MAP),* another variety of open systems interconnection, in this case developed initially by the General Motors Company as the result of the realisation that the cost of connecting unique vendors' solutions was becoming as great as the cost of the hardware itself. In this case there are a number of options for the physical medium, such as broadband, token ring and fibre optic cable.

It is also permissible to use one of the channels of a broadband system as a baseband for TOP, and it is feasible to communicate MAP on a baseband channel, although this practice is frowned upon by the originators of MAP.

The application programs for MAP are carried by the *manufacturing message service (MMS),* the upper layer of the seven-layer protocol stack. This consists of a core service and a number of companion standards for specialised applications, such as NC machines, robots and *programmable logic controllers.* The companion standards can add capability to the MMS core but in practice most of them merely modify the way the core service operates for their application area, and incorporate the appropriate specialised terminology for the field concerned. The standard documents for the MMS core and its companion standards are written in an unambiguous language (*abstract syntax notation, ASN.1*), which must be studied to determine the precise way any MMS function will work.

The lower layers of the communications stack provide error-checked data transmission to ensure the integrity of data, which is particularly important for Robot and NC applications for safety reasons.

Process Control

There is no clear recommendation for the remaining area; that of process control communications. Although a specification for an international standard for interconnecting remote sensors (known as *Fieldbus*) has been debated in the international standards forum for some time, it has not so far been possible to reach a consensus due to the existence of competing nationally-preferred contenders. As a result, there is a proliferation of proprietary solutions in this area, which is one of the most demanding in terms of the environmental conditions (harsh conditions with intrinsic safety requirements) and in the speed of response required.

For each of these areas the smaller company may find a more cost-effective unique solution for its requirements. However, the cost of the standard solutions are reducing as they can be produced in integrated circuit form rather than as software and hard-wired electronics, with the appropriate economies of scale.

Architectural Considerations

Since it is not possible to connect the TOP and MAP communications directly, owing to their use of incompatible varieties of the OSI protocol, it is useful to establish the basic architecture requirement by reference to the main functions to be included in the communications. For example, is the company predominantly design-oriented, with no manufacturing requirement other than prototype production? If so, only TOP will be required; or, is manufacturing the main concern, with a limited amount of CAD for tool design? In this case, the company can achieve an effective functionality with MAP alone. The following section explores in more detail the needs of a company with extensive activities in both design and manufacturing.

Case Study

The requirement for CIM information, from design and commercial areas to manufacturing generally, centres on the availability of bill of materials information and component part programs from the CAD/CAM system; together with details of commercial orders which will be translated into day-to-day production schedules. These schedules will also be used to determine which machines the part programs will be routed to, so there are solid reasons why the TOP network from design and commercial areas should have its principal channel through to the manufacturing control environment. Similarly, for the minute-to-minute operations, part programs can be down-loaded to the machines and routings passed to the flexible transport system using a MAP communications broadband cable around the factory. The manufacturing control function can thus act as a data exchange between the MAP and TOP networks, receiving files from one system and passing data to the other at the appropriate time, without any need for direct communication between the two systems. Equipment with inbuilt computing capability attached to the MAP network can also be used to filter information from local real-time data acquisition networks and pass exception information to the appropriate decision-making level without the need for direct communication between the two networks.

8

Standards Development

Introduction

It is one of the characteristics of the information technology field that the use of internationally accepted standards is regarded as a necessity by both user and vendors to permit successful multi-vendor installations, whilst the development of systems and hardware are proceeding too quickly to enable any individual standard to remain unchanged for any period of time without imposing unacceptable constraints on vendors' own product development.

International Standards

In these circumstances the traditional role of the standards organisations, that of documenting and codifying a stable consensus, usually already a demonstrable industrial practice, is no longer sufficient. It has been necessary for those involved in IT standards work, nationally and internationally, to take part in the development in new approaches which will be acceptable to the majority of systems vendors.

A consequence of this, and the enormous amount of time required for these tasks from unpaid industry experts, has been the pooling of individual national contributions to the development of international standards. These standards, when agreed, have an impact on all systems of whatever size, particularly those affecting communications and systems architecture. If these developments are seen as unsuitable for the needs of any particular industry sector, there is the mechanism for that sector to be represented on the relevant national standards working group and through them, the international activity. To enable potential participants in this process to become more aware of the existing activities, the areas most relevant to CIM are outline in Table 8.1. In addition, there is a regional standard (European) deriving from the Esprit CIM/OSA Project (open system architecture for computer integrated manufacture), which gives a standard approach for enterprise modelling and a means of progressing from generic to specific solutions.

	Industry	National	International
CAD/CAM data transfer	DXF	SET,VDA	IGES, STEP
CADMAT data transfer	EDIF, VHDL		
Communications			
Office			TOP
Shop-floor			MAP, Fieldbus
Inter-company			X.25, X.400

Table 8.1

9

Investment Considerations

Introduction

One of the most problematic aspects of the adoption of computer integrated manufacturing for a company of any size, has been the question of financial justification of the investment required. For the earlier examples, although numerous case studies can be found in journals and published conference papers, most decision making was conducted more as a matter of faith that the route was correct for the strategic needs of the company, rather than a strongly-argued financial justification. Partly, this was because the financial modelling techniques commonly used for corporate investment decisions were rather rudimentary mainly using return on investment or payback period methods over short intervals. This did not give an adequate demonstration of the strategic advantages, such as rapid prototype production by CAD/CAM or the avoidance of expensive retooling at the end of the first product lifecycle (which was made possible by the use of flexible and adaptable machining systems). These benefits, when recognised, would generally be considered as 'intangible' and thus dismissed from the conventional main investment analysis calculation, as being regarded as a bonus which might influence a decision between alternatives, but only if all other capital costs and running cost predictions were equal.

Although the costs of the computing side of CIM has reduced substantially, the costs of the physical embodiment, such as metal-cutting machinery and robotics, has not. For the smaller company, therefore, the risks associated with this heavy investment are at least as great, in proportion, as those of the larger companies who were the pioneers of this technology.

Fortunately in the intervening years, there have also been developments in low-cost financial modellers. These can assist by more carefully evaluating the relevant costs and their impact on the life-time economic performance of any element of the company-wide CIM implementation. Indeed, the financial modeller and its simpler cousin, the spread-sheet, were probably responsible for the widespread adoption of the desk top computer among the business community more than any other software package, and thus paved the way for the other low-cost software which makes CIM feasible for the smaller company today.

A review of the available methods of financial justification discussed below will put the suggested approach in context and perhaps help in providing bridges for including cost information already available in the company, but prepared on a different basis.

Commonly, cost information is recorded for financial accounting purposes and needs to be dealt with rather differently if it is to be used for investment analysis purposes. In particular, factors such as overhead costs, which will generally need to be evaluated in detail and included individually, as the basis for allocation or apportionment is likely to change substantially.

Traditional Methods of Evaluation

Return on Investment (RoI)

This method uses an evaluation of the average annual income generated during the life of the project, expressed as a percentage of the capital cost. This method lost favour amongst the accounting and finance profession because it did not take into account the time value of money, since income in each year of the project is considered of equal value. In practice, money received in earlier years is worth more than that from later years as it can be invested and gain interest. The method is also unsatisfactory from the engineering and business strategy point of view because it tends to assume a steady-state operation rather than the continual response to market needs, which is the reality for most companies. It is thus likely to favour fixed investment in a long-term product rather than the more flexible facility which can be easily retooled for future product needs.

Payback Period

In contrast to the Return on Investment method, this approach evaluates the number of years before which the cumulative income from the project exceeds the original capital cost. This clearly excludes the benefits of longer-term advantages and often, if used for comparison purposes, favours a project with high early returns rather than one with a higher overall earning potential but which requires more time to establish. Like the *RoI* method, however, it takes no account of the time value of money.

Although the payback period method is judged unsatisfactory as the principal means of comparison for competing investment projects, it nevertheless has a legitimate place as a quick measure of the time by which any project is at risk, before earnings exceed outlay.

Recommended Methods of Evaluation

Having dismissed the two traditional evaluation methods, it remains to discuss in more detail the two methods which are recommended. Both of these take account of the time value of money by applying a successively higher rate of discount to the earnings, generated by the project in each of the years following the year of implementation. This gives the value of the income in the equivalent of money in the bank at the present day, or present value. They are both therefore known as *discounted cash flow (DCF)* methods, although this title is sometimes used for the first of these methods, otherwise known as the *internal rate of return* method.

Internal Rate of Return (IRR)

The internal rate of return method of evaluation is performed by first evaluating the costs incurred and incomes generated in each year during the lifetime of the project. Successively higher discount factors are then applied to these annual cash flows until the net present value of the project (income minus/capital cost plus running costs) is equal to zero. The discount rate which has been used to generate the discount factors for each year of the project is termed the *internal rate of return.*

A disadvantage of this method is that when comparing two high-earning projects, the discount rate which must be applied to reduce the present value to zero may be unrealistically high and will not represent the rate of interest which the company could expect to achieve from normal investment. In these circumstances, this discount rate is likely to cause distortions in the evaluation by underestimating the value of income received in later years.

When applied to individual projects (perhaps a stand-alone cost saving project), the rate of IRR discount which should be exceeded before the project can be considered for inclusion in the investment budget, is known as the *hurdle rate*. It is not suggested that this approach is applied in a strict manner to CIM projects as there would be a danger that a small project, which may not be particularly cost-effective but is of strategic importance for a number of high-earning projects, could be excluded.

Net Present Value

This approach again uses discounted cash flow but, rather than apply a number of different rates, the company first decides which is the appropriate interest rate based on money market predictions and its own circumstances as net borrower or lender. This rate is then used to generate the discount factors for each year of the project. The net value of the cash flow, positive or negative, in each year over the lifetime of the project is used as a measure of the worth of the project.

This method has the advantage that realistic forecast incomes are used in each year, so it is possible to include the minor costs and benefits likely to result from implementation of the project. It becomes worthwhile to spend more effort in evaluating some of the more indirect spin-off benefits of the project, or 'making the intangible tangible'. This is the preferred method.

10

Future Possibilities

Introduction

Although it is never possible to predict with certainty the pattern of future developments, let alone the extent of their adoption, it is possible to give some indication of the direction in which further enhancements of existing technologies are likely. For more extensive forecasts, the reader is invited to read the numerous Delphi forecasts compiled from inputs by industry experts and managers, or to invest in a personal crystal sphere!

Manufacturing Control

It was mentioned in Chapter 4 that a feature of MRP II systems was some degree of simulation of possible eventualities from different courses of action, and that OPT software was centred on a more detailed simulation of factory operation. The future course of development is likely to involve developments of the more sophisticated discrete event simulators, incorporating detailed models of the particular factory or commercial operation, linked to expert systems, initially for decision support to the human manager. By incorporating real-time communications from a multiplicity of data input sources (mainly automatic sensors), it will be feasible to quickly calculate a number of possible outcomes before an irrevocable decision must be made.

CAD/CAM

The extended adoption of geometric solid modellers is likely to be overtaken by the move to feature-based modelling, probably in conjunction with expert systems to carry out detail design based on formal design specification. The mathematical basis of the CAD system will permit the assessment of tolerance aspects and the associated manufacturing capability requirements. Rapid design will be complemented by new fast prototype processes such as the production of three-dimensional solids from solidification of liquid resin, cured by laser beam manipulation. It is also likely that the results of stress analysis calculations will be displayed more effectively by laser holography.

Communications

After the development of a more extensive application of fibre optic transmission for multi-purpose factory communications, a number of functions will probably be carried out by radio and infra-red transmission, avoiding the need for cabling.

Access to wide area networks with engineering database facilities will have a great effect on the need to store data locally, thus minimising the mass storage requirements of the selected computer hardware, for instance, the availability of catalogues of standard parts over the network will eliminate the need to store and update the large number of standard parts which the typical engineering designer needs to select from. Until this capability is available, the standard catalogues of parts will be available from the major suppliers in disc form, encouraging the retention of existing industry standard data formats even when higher speed and density technology becomes available.

User Interface

The general development of a standardised approach to user interface requirements among a range of software applications (a common 'look and feel'), is progressing amongst the IT community and may be expected to be adopted by all vendors for the commercial elements of CIM software, such as sales order processing and stock control. However, this is not expected in the CAD/CAM area. In addition to the differences between offerings from the software vendors, the user interface is generally customised with the use of specialised menu format for each specialist task requirement from concept design through to maintenance handbook illustrations.

The current research activities into virtual reality user interfaces, using stereo vision headsets for visualisation of three-dimensional environment and a tactile feedback glove for interaction with the environments, is likely to enhance these differences and encourage the development of more specialised interaction facilities. For example, the stereo headset has already proved its value in giving the realistic impression of a simulated walk through a detailed three-dimensional model of a major architectural project, such as a shopping centre.

The addition of tactile feedback gives the capability of providing an intuitive control of the manipulation of a three-dimensional solid as a sculptor would shape a piece of clay, allowing greater scope for artistic expression in industrial design than is possible with the conventional user interface.

Glossary

Application Layer — *see also* OSI. The application layer may also incorporate additional software to handle data interchange facilities, for example, the Initial Graphics Exchange Specification (IGES) and Electronic Design Interchange Format (EDIF).

AGV — (Automated Guided Vehicle.) A mobile robotic device.

ATE — (Automated Test Equipment.) Equipment which can be programmed to carry out a variety of measurements on the key functional parameters of an assembly of electronic components.

BTR — (Behind the Tape Reader interface.) A form of interface used to adapt Numerical Control and Computer Numerical Control equipment, without communications facilities, to the requirements of Direct or Distributed Numerical Control. An interface is installed alongside the machine to handle the communications and supply the numerical data to the machine interface normally used for the part program tape reader.

CAD — (Computer Aided Design.) The use of a computer for calculations or geometric modelling of mechanical or electrical engineering components and assemblies. The geometry is defined by a dimensional model held as a data file in the computer.

CAD/CAM (Computer Aided Design and Manufacturing.) The use of Computer Aided Design for the representation of the geometry of a mechanical component, linked with computer assisted part programming to generate a data file for the numerically controlled machine tool which will be used to produce the part.

CADMAT (Computer Aided Design, Manufacturing and Test.) The use of Computer Aided Design for the representation of the functional description and definition of electronic components and assemblies. The data generated can be used to control manufacturing equipment and Automated Test Equipment.

CAD (Computer Aided Draughting.) The use of a computer to assist the detail draughtsman in the production of engineering drawings which give a visual representation of the component or assembly concerned. The drawing is then annotated with toleranced dimensions to define the geometry.

CAE (or MCAE) (Computer Aided Engineering.) The use of computer aided design for geometric modelling of mechanical engineering components and assemblies using analysis techniques such as Finite Element, Finite Difference and Boundary Elements to predict their behaviour in service.

CAPM (Computer Aided Production Management.) The use of a computer to assist the management of the logistics aspects of the manufacturing process, such as order processing, scheduling and inventory management.

CAPP (Computer Aided Process Planning.) The use of a computer to assist the selection of the appropriate method of manufacturing for an engineering component.

CIM (Computer Integrated Manufacture.) The use of a central computer or networked distributed computers to integrate the functions of a manufacturing organisation, particularly those concerning marketing, design and manufacture.

CNC (Computer Numerical Control.) The control of a manufacturing operation by equipment which includes a stored-program computer to adjust the sequence of operations and the dimensions of the components produced in response to numerical data (the part program). The data may be fed to the computer as the operation progresses or stored as a data file in the computer memory for repetition, as required.

DNC (Direct Numerical Control.) The control of a manufacturing operation by means of numerical data in which the data is communicated from a central storage file as the operation progresses. This requires communication links from the equipment to the central storage computer to be active whenever the equipment is in operation.

DNC (Distributed Numerical Control.) The control of a manufacturing operation using a stored-program computer for equipment functions and a central computer for data storage. The numerical data can be communicated as a data file to the equipment and this data (the part program) can be stored at the equipment for repeated usage with no further communication taking place until a different set of data is required.

EDI (Electronic Data Interchange.) The communication of commercial business data such as orders and invoices by electronic means. A number of formats are in existence to meet the requirements of specialised industry sectors.

EDIF (Electronic Design Interchange Format.) An industry standard neutral format for the exchange of design data for electronic devices such as integrated circuits and printed circuit boards.

MCAE *See* CAE.

NC (Numerical Control.) The control of a manufacturing operation by means of numerical data which is fed to the equipment as the operation progresses. The equipment may include purely mechanical devices or a combination of mechanical and hard-wired electronics control systems.

OSI (Open Systems Interconnection.) A framework of seven layers of inter-changeable elements intended to standardise the interfaces required in communications networks. The layers may be grouped into two subsets for convenience:

A-subset of layers - Application (7), Presentation (6), Session (5)

T-subset of layers - Transport (4), Network (3), Link (2) and Physical (1).

For some purposes a preferred set of layers, known as an Industry Standard Profile (ISP) can be specified. Examples are: Manufacturing Automated Protocol (MAP), and Technical and Office Protocol (TOP).

PLC (Programmable Logic Controller.) A sequencing and control device, generally based on a microprocessor, which features rugged construction and protection of digital and analog inputs and outputs to suit harsh industrial environmental conditions. The devices were developed to replace mechanical relay logic control and are often programmed using a graphical display which represents relay coils and contacts.

RAM (Random Access Memory or Read-Write Memory.) The main storage for program and data when a computer is in operation. This is now generally in the form of semi-conductor memory chips which are erased whenever the computer is switched off, but was originally in the form of arrays of small ferrite magnetic cores and is thus sometimes called Core Store.

ROM (Read Only Memory.) Storage for computer programs or data which is not changed during the normal operation of the computer. In addition to semiconductor chips which are programmed during manufacture, there is a variety of chips which can be programmed subsequently, including:

PROM - may be programmed electrically once only.

EPROM - may be programmed electrically and then erased for reuse, generally by exposing the chip to ultraviolet light.

EAROM - may be electrically altered from one data set to another.

Bibliography

CAD/CAM, M.P. Groover and E.W. Zimmer, Prentice-Hall, 1984

Computer-Integrated Manufacturing, Alan Weatherall, Butterworths, 1988

Design Rules for CIM Systems, R.W. Yeomans et al., North-Holland, 1985

The Goal, E.M. Goldratt and J. Cox, Scheduling Technology Group, 1987

What is MAP?, Colin Pye, NCC Blackwell, 1988

Index